HYDROPOWER ECONOMICS

***Recent titles in the* INTERNATIONAL SERIES IN OPERATIONS RESEARCH & MANAGEMENT SCIENCE**
Frederick S. Hillier, Series Editor, *Stanford University*

Sethi, Yan & Zhang/ *INVENTORY AND SUPPLY CHAIN MANAGEMENT WITH FORECAST UPDATES*
Cox/ *QUANTITATIVE HEALTH RISK ANALYSIS METHODS: Modeling the Human Health Impacts of Antibiotics Used in Food Animals*
Ching & Ng/ *MARKOV CHAINS: Models, Algorithms and Applications*
Li & Sun/ *NONLINEAR INTEGER PROGRAMMING*
Kaliszewski/ *SOFT COMPUTING FOR COMPLEX MULTIPLE CRITERIA DECISION MAKING*
Bouyssou et al/ *EVALUATION AND DECISION MODELS WITH MULTIPLE CRITERIA: Stepping stones for the analyst*
Blecker & Friedrich/ *MASS CUSTOMIZATION: Challenges and Solutions*
Appa, Pitsoulis & Williams/ *HANDBOOK ON MODELLING FOR DISCRETE OPTIMIZATION*
Herrmann/ *HANDBOOK OF PRODUCTION SCHEDULING*
Axsäter/ *INVENTORY CONTROL, 2nd Ed.*
Hall/ *PATIENT FLOW: Reducing Delay in Healthcare Delivery*
Józefowska & Węglarz/ *PERSPECTIVES IN MODERN PROJECT SCHEDULING*
Tian & Zhang/ *VACATION QUEUEING MODELS: Theory and Applications*
Yan, Yin & Zhang/ *STOCHASTIC PROCESSES, OPTIMIZATION, AND CONTROL THEORY APPLICATIONS IN FINANCIAL ENGINEERING, QUEUEING NETWORKS, AND MANUFACTURING SYSTEMS*
Saaty & Vargas/ *DECISION MAKING WITH THE ANALYTIC NETWORK PROCESS: Economic, Political, Social & Technological Applications w. Benefits, Opportunities, Costs & Risks*
Yu/ *TECHNOLOGY PORTFOLIO PLANNING AND MANAGEMENT: Practical Concepts and Tools*
Kandiller/ *PRINCIPLES OF MATHEMATICS IN OPERATIONS RESEARCH*
Lee & Lee/ *BUILDING SUPPLY CHAIN EXCELLENCE IN EMERGING ECONOMIES*
Weintraub/ *MANAGEMENT OF NATURAL RESOURCES: A Handbook of Operations Research Models, Algorithms, and Implementations*
Hooker/ *INTEGRATED METHODS FOR OPTIMIZATION*
Dawande et al/ *THROUGHPUT OPTIMIZATION IN ROBOTIC CELLS*
Friesz/ *NETWORK SCIENCE, NONLINEAR SCIENCE and INFRASTRUCTURE SYSTEMS*
Cai, Sha & Wong/ *TIME-VARYING NETWORK OPTIMIZATION*
Mamon & Elliott/ *HIDDEN MARKOV MODELS IN FINANCE*
del Castillo/ *PROCESS OPTIMIZATION: A Statistical Approach*
Józefowska/*JUST-IN-TIME SCHEDULING: Models & Algorithms for Computer & Manufacturing Systems*
Yu, Wang & Lai/ *FOREIGN-EXCHANGE-RATE FORECASTING WITH ARTIFICIAL NEURAL NETWORKS*
Beyer et al/ *MARKOVIAN DEMAND INVENTORY MODELS*
Shi & Olafsson/ *NESTED PARTITIONS OPTIMIZATION: Methodology And Applications*
Samaniego/ *SYSTEM SIGNATURES AND THEIR APPLICATIONS IN ENGINEERING RELIABILITY*
Kleijnen/ *DESIGN AND ANALYSIS OF SIMULATION EXPERIMENTS*

**** A list of the early publications in the series is at the end of the book ****

HYDROPOWER ECONOMICS

by

Finn R. Førsund

Springer

Finn R. Førsund
University of Oslo
Blindern, Norway

Series Editor:
Fred Hillier
Stanford University
Stanford, CA, USA

ISBN-13: 978-1-4419-4455-9
ISBN-13: 978-0-387-73027-1 (e-book)

Printed on acid-free paper.

9 8 7 6 5 4 3 2 1

springer.com

Preface

The state organisation responsible for coordinating the hydropower electricity system in Norway ("Samkjøringen") contacted me in 1990 about the advanced plan for deregulating the electricity system, separating generation, transmission, and distribution and introducing a wholesale market for electricity. It was felt that insights about the fundamental nature of running an electricity system based on hydropower was somewhat lacking within the team of academic economists engaged to write background reports by the Oil and Energy ministry responsible for driving the reform of the electricity system.

When talking to engineers I was fascinated by the world of electricity, with its physical laws and weird concepts such as reactive power and electric phase angles. Externalities of hydraulic interdependence between river-based power stations and highly fluctuating loss and congestion externalities involved in a meshed transmission network had to be recognised. Furthermore, capturing all these elements required advanced mathematical methods of dynamic programming in a stochastic environment. My conclusion was that a market design that neglected these aspects did it at its own peril. I predicted volatile prices coming out of a competition between producers facing zero short-run variable costs and problems with investments coming forth sufficiently from a social perspective. However, I can safely say that my report had no impact whatsoever on the Norwegian electricity reform of 1991, that must be regarded, not the least by me, as being highly successful.

The main result of my report was that I became fascinated with hydropower economics and started to lecture on the topic at my department of economics at University of Oslo. However, I had difficulties finding texts that were suitable for economists. The field is well developed within engineering, but aspects of economics of hydropower were not so easy to come by. My great inspiration has been two papers by Hveding (1967, 1968). He was an engineer and general director of the electricity regulation body in Norway (NVE), and followed up the great tradition of engineers at Electricité de France (EDF) of writing exciting stuff that economists could also appreciate.

The Nordic Council research project Energy and Society, headed by Torstein Bye, gave me opportunities several times over the years to present my ideas at Nordic workshops, and made it possible to develop these ideas on an extended visit to Iceland.

It is the generous support by Norway's biggest hydropower producer, Statkraft that finally made it possible for me to develop my material into a book. Statkraft bought me free from my teaching and administrative obligation at my department for half a year. I especially thank Geir Holler for his trust in me, he also took my course in natural resources when I developed the hydropower theme, and Kjell Berger for providing me with data and reading parts of the manuscript and offering sobering comments.

I will also like to thank Tor Arnt Johnsen at NVE for encouraging me to carry out the project and helping me initially seeking finance. My colleague Atle Seierstad generously used his time to advise me on the use of mathematics, and I owe Kjell Arne Brekke warm thanks for enlightening me on uncertainty. Torstein Bye, Stein-Erik Fleten, Richard Green, Petter Vegard Hansen and Lennart Hjalmarsson have read parts of the manuscript and offered valuable comments. They are in no way responsible for remaining deficiencies.

I was fortunate to become a visiting fellow at International Centre for Economic Research (ICER) in Torino, which provided me with the perfect environment to write a book during spring 2006. I will like to thank Alessandra Calosso at ICER for excellent assistance, not the least in times of crisis, such as breakdown of my PC hard disk.

When Springer provided me with a 25-page manual on how to construct the special layout for the book, I knew I was in serious trouble with managing the last hurdle. Fortunately Marius Østli came to my rescue and did an excellent job of converting my manuscript in Word into the Springer layout standard. In addition he has provided solid support making the finishing touches to the manuscript.

Last, but not least, I want to thank Marisa for her support, inspiration, and understanding.

Finn R. Førsund
Torino, 20 June 2007

Contents

List of Common Variables

Variable	Definition	Unit of measurement
T	Planning horizon	Periods
R_t	Amount in the reservoir at the end of t	m^3 (kWh)
$\overline{R}$	Reservoir capacity	m^3 (kWh)
R^T	Minimum level in reservoir	m^3 (kWh)
W	Total available inflow	kWh (m^3)
λ	Shadow price on stored water; water value	Money / kWh (m^3)
w_t	Inflow of water during t	m^3 (kWh)
w_t^R	Inflow of water during t to a river plant	m^3 (kWh)
r_t	Release of water during t	m^3
r_t^u	Maximal ramping up during t	m^3 (kWh)
r_t^d	Maximal ramping down during t	m^3 (kWh)
r_t^g	Water to group g in period t	m^3
a	Fabrication coefficient	m^3 / kWh
e_t^H	Electricity production from regulated hydro during t	kWh
e_t^R	Electricity production from unregulated river during t	kWh
e_t^W	Electricity production from wind power during t	kWh
e_t^L	Loss of electricity during t	kWh
e_{jt}^{ru}	Maximal ramping up of plant j during t	kWh
e_{jt}^{rd}	Maximal ramping down of plant j during t	kWh
$\overline{e}^H$	Production capacity of hydro	kWh
$\underline{e}_{jt}^H$	Minimum production constraint of plant j during t	kWh
e_t^{XI}	Export/import during t	kWh

Variable	Definition	Unit of measurement
$\overline{e}^{XI}$	Export/import capacity	kWh
e_t^{Th}	Electricity production from thermal capacities during t	kWh
$\overline{e}^{Th}$	Production capacity of thermal	kWh
$c(e_t^{Th})$	Thermal cost function	Money
U_t	Social utility of electricity consumption during period t	Utils (money)
p_t	Social price of electricity period t	Money / kWh
p_t^{XI}	Export/import price period t	Money / kWh
x_t	Consumption of electricity during period t	kWh
$x_t^{\max}$	Highest electricity consumption period t	kWh
b_{st}	Flow on line s during t	kWh
$\overline{b}_s$	Capacity limit on line s	kWh
$p_t(.)$	Demand function for electricity on price form period t	Money / kWh
β_t	Discount factor period t	
$S(R_T)$	Scrap value function for period T	Money
N	Number of plants	
N^I	Number of independent plants	
N^C	Number of coupled plants	
N^R	Number of unregulated plants	
L_t	Labour input during t	Hours
E_t	Energy input during t	kWh
$z_{ip,t}$	Emissions plant i of pollutant p during t	kg
$\overline{z}_{ip,t}$	Emission limit plant i of pollutant p during t	kg
z_{it}	Emission plant i during t	kg
$\overline{z}_t$	Emission limit during t	kg
$\breve{\eta}_t$	Demand flexibility period t	

Chapter 1. Introduction

Background

Domestic pricing of hydropower was for many years an area of direct political control in Norway. After the parliament restricted both domestic and foreign private ownership of waterfalls for hydropower development soon after Norway became an independent country again in 1905, the public sector has been the dominant provider of electricity, at present owning almost 90% of the hydropower capacity. At the municipal level, providing electricity for general purpose consumption, the pricing policy was based on average cost pricing, while the state-owned power stations, feeding the national grid, delivered power mainly to energy-intensive industries like aluminium, ferro alloys, and pulp and paper to very favourable prices. Greenhouse activities are also favourably treated as part of the protective agricultural policy pursued by Norway. The cheap electricity was a main localisation factor for primary aluminium industry because all other raw materials, like aluminium oxide, are imported, and although part of the technology was developed in Norway (the Søderberg anode), the technology is now international. The cheap electricity policy may have been appropriate when considering electricity supply in autarky, although a statement from an influential former prime minister (educated as an economist) may cast some doubt on the quality of the social cost-benefit analysis behind the policy that had widespread political acceptance:

> If one wants cheap electricity one must build so much capacity that there is enough electricity at the price one wants (Willoch, 1985).

Deregulation of the electricity industry came on the political agenda in many countries around the world in the late 1980s and early 1990s. The creation of a wholesale market as a day-ahead last price auction in England in 1990 (Newbery, 2005) started a process toward similar deregulation in Europe that is still taking place. Besides the political aspect of a policy of privatisation there was an economic rationale of competition driving down production costs and price. Production took place in numerous units that in principle could compete, although transmission was an area of natural monopoly.

Norway followed up England's type of deregulation by setting up a similar competitive wholesale market in 1991. However, while the production in England was based on 63 conventional coal-based thermal units and 12 nuclear plants organised into only three companies (plus a modest pumped-storage capacity) (Newbery, 2005), the production in Norway was based on hydropower supplied by over 600 plants. The operation and management of these plants had mainly been seen as tasks for engineers only. Within the national grid the electricity regulator (NVE) used system analysis to coordinate the management of the reservoirs of water for the total system in such a way that in principle total demand was satisfied in the cheapest way, observing the requirement of supply by municipal hydropower plants. The electricity regulator was also responsible for watching over the energy balance and keeping the politicians informed and for planning capacity expansion.

Economists in Norway had for many years been critical both of the political pricing policy of electricity in Norway, resulting in prices varying both regionally and between different user groups, and of the criteria used for expanding capacity resulting in too rapid expansion and without environmental considerations taken properly into account.[1] The period of expanding the hydropower capacity had in fact come to an end due to a lack of reasonably profitable hydropower projects without a strong opposition from environmental interest groups when deregulation came on the political agenda. The transition from central coordination and control to a market-based wholesale competition between producers went remarkably smoothly. It is also remarkable that the introduction of a market took place with almost 90% of the production capacity being publicly owned (35% state and 55% municipal ownership). However, it is not easy to find evidence of cost reductions on the production side that was used as one of the arguments for introducing a market in generation. Hydropower is in fact run with negligible variable costs. The people employed and the maintenance costs can be regarded as fixed costs independent of variations in production, but related more to the production capacity. In view of the promise of reducing generating costs it is remarkable that the only study I know about costs has been done as a master's thesis by one of my students (Lien, 2006). The study found a modest decrease in fixed costs over time since deregulation and a systematic substitution of permanent employees by outsourcing.

---

[1] See the discussion by leading economists in NOU (1979)

One result of the market reform is that prices to households and the general commercial sector have for the most part evened out between regions. Consumers have a free choice of supplier and can switch without costs. The prices have also been on a rather low level internationally. However, this is due mainly to the excess capacity in the system before deregulation, and prices are on their way up now. The power-intensive industries managed to hang on to their cheap electricity contracts forced upon the state power company by the politicians. The contracts expire from 2005 to 2011.

The intention of the new electricity regime was that market actors themselves should undertake investment in new capacity. However, so far the investments have been negligible. This is probably mainly a reflection of the extent of over-investment previously, but also the benefit of extending the market using existing capacity more efficiently.

One remarkable achievement of the market reform on the wholesale side is that Norway pioneered trade over borders and in fact created the first integrated international market, Nord Pool, in electricity together with Sweden in 1996. Later Finland and Denmark joined Nord Pool. Although the technical possibilities for trade of electricity with neighbouring countries like Sweden and Denmark had been there for a long time before deregulation, Norwegian politicians followed a principle of not allowing trade with "firm power," i.e., the amount of hydropower electricity one would expect to produce in 9 out of 10 years.[2] But bilateral trade in "occasional power," i.e., power in years with unexpectedly high rainfall, was developed with Sweden and Denmark over many years, especially for use in industrial boilers that could easily switch between primary energy sources. These trades were a forerunner of the Nord Pool market developed in the 1990s. International trade now takes place between many European countries on a bilateral basis, e.g., France-England, France-Italy (Italy is importing about 20% of its electricity), etc. The energy policy of the European Union is encouraging a gradual expansion of cross-border trading and integration of national electricity markets (Jamasb and Pollitt, 2005).

[2] In NOU (1979) it was argued that the alternative cost of Norwegian hydropower was the price that could be obtained on the export market, presumably much more than the Norwegian power-intensive industry was enjoying. Notice that this argument was used over a decade before the self-sufficiency policy was abandoned.

The purpose of this book

According to information provided by the International Hydropower Association (www.hydropower.org) 20% of the world's electricity is produced by hydropower and one third of all countries in the world depend on hydropower for over 50% of their electricity generation (in 2001). Norway's electricity production is based 99% on hydropower. Other countries that also have a high share of hydropower are Brazil (90%), Iceland (88%), New Zealand (65%), Austria (70%), Canada (62%) and Sweden (40%). The United States has a 7% share of hydropower, but is the biggest producer next to Canada. The other top producers are Brazil, China, Russia, and then Norway as the sixth biggest producer worldwide. But we should also pay attention to the regional importance of hydropower. In some western US states hydropower is more important, as is also the case for, e.g., the province of Quebec in Canada. Because of this worldwide use of hydropower it is important to understand how to operate hydropower and the interaction of hydropower with other producing technologies of electricity.

The main purpose of this book is to provide qualitative economic analyses of how to utilise stored water in a hydropower system, i.e., problems of current management with fixed generating capacities. This problem is a dynamic one because water used today to generate electric power may alternatively be used tomorrow. Understanding and evaluating today's deregulation requires a sufficient background in the theory of optimal use of hydro and thermal power by economists, engineers, and regulators involved in managing the electricity system.

The problem of optimal investment is not addressed in this book. This is in itself a major undertaking. However, in order to solve this problem successfully the management problem of optimal use of stored water, given the production capacity, must also be solved simultaneously.

Hydropower is a field within engineering. But, as remarked by Edwards (2003) in the motivation for his book on the subject, economic analyses are found scattered around in journal articles and are not satisfactorily treated in a book addressed to economists. However, the need for a comprehensive text still exists, one reason being that Edwards (2003) focuses exclusively on small-scale systems of power stations located along rivers run by a local authority, and has a considerably more limited scope than the present book.

The economics of hydro production with reservoir was discussed early in the operations research and economics literature (Little, 1955; Koopmans, 1957; Morlat, 1964[3]), but the topic is a typical engineering one (a well-known textbook is Wood and Wollenberg, 1984). In Norway a national central coordination system for hydropower production was established after World War II based on an understanding of how the total system was to be operated (see Hveding (1967, 1968).[4] This approach has been refined and developed into a central model tool for Norway and later the Nord Pool area (Johannesen and Flatabø, 1989; Haugstad et al., 1990; Gjelsvik et al., 1992; Wangensteen, 2007). The highly simplified approach taken in this book is based on the approach used in Førsund (1994) (for related model concepts see also Bushnell, 2003; Crampes and Moreaux, 2001; Johnsen, 2001; von der Fehr and Sandsbråten, 1997; and Scott and Read, 1996). A comprehensive literature review is not offered, and only papers of importance for developing the analyses in this book are referred to.

The main inspiration for the present book has been the articles by Hveding (1967, 1968), as will be evident by the references in the relevant chapters. The distinctive feature of this book is to provide a social planning perspective on optimal use of water, which is a prerequisite for understanding and evaluating the newly established electricity markets.

The dynamic nature of hydropower production, the high number of units involved, and the inherent stochastic nature of key variables like inflow of water make optimisation problems quite difficult technically to solve. In the engineering literature, based on the Bellman (1957) approach to dynamic programming, sophisticated stochastic dynamic programming models are used and solution algorithms developed for real-life data and numerical solutions provided. I will try to use a much more simplified mathematical approach suited to obtain qualitative conclusions. As to the choice of theoretical modelling, standard nonlinear programming models for discrete time are used and the Kuhn – Tucker conditions employed extensively for qualitative interpretations. This choice of modelling cannot be better motivated than expressed by the following quotes from Baumol (1972):[5]

---

[3] In France there were early studies from the 1940s and 1950s, especially by people connected to Electricité de France; see the references in Morlat. See also review of French contributions in the Introduction in Nelson (1964), and for translations into English of other French papers.

[4] Hveding was the general director of the electricity regulator, the Norwegian Water and Energy Directorate (NVE) from 1968 to 1975.

[5] I am indebted to my friend Professor Mikulas Luptacik for bringing these quotes to my attention.

> ....economists have used them [the Kuhn – Tucker conditions] primarily to deal with more general qualitative problems. That is, the conditions can be used to derive *general* conclusions about the nature of the solutions, ... (p. 165).
> ....the Kuhn – Tucker conditions may perhaps constitute the most powerful single weapon provided to economics theory by mathematical programming (p. 165).
> ....It is therefore a manifestation of the very great power of the Kuhn – Tucker analysis that it does permit us to arrive at general qualitative conclusions about the behavior of the solutions to nonnumerical problems (pp. 165-166).

In order to strengthen the understanding of the basic nature of the solution to the dynamic hydropower problem, graphical illustrations are developed and used extensively. Two periods often suffice to capture the main understanding of a dynamic problem, and it is therefore possible to illustrate such an understanding. A special graphical presentation, termed a *bathtub diagram*, is developed.

The plan of the book is the following: the rest of Chapter 1 very briefly covers the nature of electricity involving the concepts power and energy and the instantaneous equilibrium between supply and demand. Load-duration curves for different time units for Norway are used to illustrate concepts like peak and base load. The nature of hydropower production is introduced using a production function and presenting the fundamental water dynamics of the reservoir constraints. The environmental problems associated with hydropower are briefly summarised.

Chapter 2 presents the basic hydropower model without a reservoir constraint. Electricity consumption is evaluated by utility functions. Water is treated as a natural resource in finite supply within the planning horizon, and the Hotelling rule for pricing of a finite natural resource is derived also in the case of discounting. The case of several user groups of water is treated and the equality of (socially weighted) marginal utilities between groups and over time is established.

Chapter 3 introduces the typical constraints faced by a hydropower system. The generator capacities are aggregated into a single system with a single reservoir and analysed within a given horizon of multiple periods. A social planning model with a reservoir constraint that may become binding, showing the fundamental dynamics involved, is introduced, and economic interpretation of first-order conditions performed. The bathtub diagram is used to show two consecutive periods together. Emphasis is put on events that will lead to a change in the (social) price of electricity. The events are threat of overflow of the reservoir and emptying the reservoir. Further extensions are introducing upper limits on the production (or power) capacity, and introduction of run-of-the-river hydropower without storage and wind mills. The implications for optimal hydro management and prices are derived.

Chapter 4 models multiple generators and reservoirs within a multiple-period planning horizon. It is shown that the optimal use of multiple generators and reservoirs can facilitate considering aggregation of generators into one unit and reservoirs into one reservoir, greatly simplifying the derivation of the social solution. There is no unique solution for individual generators except that the individual reservoirs should fill up the reservoir to the limit in the same period and should be emptied in the same period. The aggregation result is called the Hveding conjecture. However, the conjecture holds only for specifying reservoir limits. When introducing also production (power) limits, the optimal solution for individual units becomes more complicated and an aggregation into a single system will only serve as an approximation. Optimality conditions involving hydraulically coupled generators are derived and consequences of environmental constraints explored.

Chapter 5 introduces thermal generators together with hydropower. The assumptions leading to merit-order aggregation of a short-run aggregate supply function are given. A special bathtub diagram is developed for a graphical presentation of the mix of hydropower and thermal capacity. For periods with the same price the same amount of thermal capacity is used, while hydropower use follows variation in demand. The mix of hydro-thermal capacity as peak load and base load is discussed. The introduction of start-up costs of thermal generators leading to the optimisation of use of thermal units is demonstrated.

Chapter 6 extends the analysis to trade between countries in the case of fixed foreign prices. The conditions under which foreign prices will be adopted as domestic prices are investigated. The consequences of constraints on transmission between countries are explored, and the case of trade between a hydro country and a thermal country with endogenous prices is studied. Both the impacts of only a total water (energy) constraint and a reservoir constraint are investigated.

A transmission network is introduced in Chapter 7 by using a highly simplified way to model loss and congestion in the network. The external effects of creating losses are brought out. Congestion of lines is introduced, but without modelling loop-flow effects. The general conclusions confirm the findings in Schweppe et al. (1988) of specific nodal prices both for generating and consumer nodes. The use of hydro reservoirs is influenced both across and over time by transmission.

Chapter 8 deals with market power. A monopolist may spill water in order to contract production in the classical way, but the general new feature in the hydropower context is the shifting of water away from relatively inelastic demand periods to use in relatively elastic demand periods when there is no spilling and the same total production is maintained. The consequences of trade, mix of hydro and thermal capacity, and a competitive fringe with thermal capacity are studied.

Chapter 9 introduces uncertainty. The implications of stochastic inflows for modelling and conclusions for pricing are explored with regard to qualitative features of the optimal social solution. The basic outcome of optimisation is a decision rule to be followed as time evolves. An important qualitative result is that prices may vary over periods even if the expected prices ex ante may all be the same. The simple reason is that the successively realised inflows may deviate from the expected levels, making continuous adjustment of prices as time evolves toward a planning horizon the optimal policy.

Some concluding comments are offered in Chapter 10 concerning lessons learned and the light they can shed on actual electricity markets and policies. It is important to realise that the theoretical modelling is based on formulating demand functions in real time. This is very seldom the case in practical market or planning-based systems. This fact, together with the externalities caused by hydrological coupling and generation of transmission losses and congestion of transmission lines, casts some doubt on the practice of appealing to the welfare theorems concerning optimality properties of market systems when using theoretical model solutions not taking these phenomena into consideration. Although investment problems have not been addressed, the values of shadow prices on various capacity constraints may serve as indications of the profitability of marginally increasing the capacities. In equilibrium, both with respect to operations and capacities, it should not be profitable with marginal increases of capacities.

Electricity

Electricity is one of the key goods in a modern economy. The nature of electricity is such that supply and demand must be in a continuous physical equilibrium. The system breaks down in a relatively short time if demand exceeds supply and vice versa. A system failure may lead to grave economic consequences if the blackout lasts too long. Recent failures of shorter duration in the United States, Canada and Europe have led to more inconvenience than serious economic damage, and the more amusing effects of more babies being born 9 months later as was the case after a period of blackout in New York some years ago.

The spatial configuration of supply and demand is important for understanding the electricity system. A transmission network for transport of electricity connects generators and consumers. There is energy loss in the form of heat in the network. Physical laws govern the flows through the networks and the energy losses. Electricity delivered to general consumers is characterised by voltage (220-240 volts in Europe, 110-120 volts in the United States) and frequency measured in Hertz (50 ± 0.1 in Europe, 60 in the United States) for alternate current. Electricity is measured as power (kW), i.e., instantaneous energy, and energy (kWh), i.e., the amount of electricity during a time period (the integral of the power over the time period in question). A central operator that secures equilibrium in continuous time usually runs the system. The equilibrium is then in power. This operator should be independent of suppliers and consumers, and may also be responsible for running the transmission network. Normal operating procedure is to take demand as given and adjust supply.

The economic notion of a price adjusting in order for demand to equal supply within a time period, e.g., the demand and supply for apples during a market day, is therefore not immediately applicable to electricity markets. However, the assumption that demand depends on price should still be useful, although one has to be more careful about distinguishing between short and long run and whether pricing is in real time or applies ex post.

Demand for electricity

The time period used in a study of the electricity system is of crucial importance for the detail by which the system is modelled. In continuous time the demand is for power, and energy will be the integral over the time periods chosen. If time is discrete it is usually assumed that the power is constant over the chosen time period. The demand can then be expressed either for power or energy.

In order to understand the variation in demand for electricity it is useful to consider the various uses. Household demand is for light, hot water, cooking, running various appliances like TV, refrigerators, washing machines, etc., and space heating. The last use represents about 31% of household electricity use in Norway, as seen in Figure 1.1. The household shares of electricity are found in Larsen and Nesbakken (2005) by conducting conditional demand analysis (CDA) based on 987 Norwegian households for 2001. The group "Other" comprises cookers, motorcar engine heaters, sauna, TV, etc. The category most easily substituted by

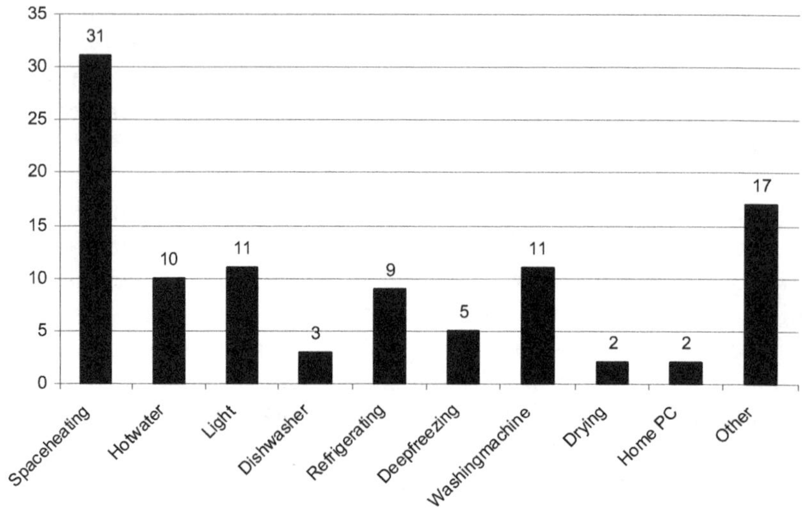

Figure 1.1. Household shares of electricity consumption in percent. Norway 2001.
Source: Larsen and Nesbakken (2005)

other energy sources is space heating. About 38% of dwellings also have other systems like oil burners, paraffin heaters, wood stoves, etc.

In industrial use, in addition to light, hot water, and office heating, there are machinery, process heat, and electrolytic processes. An interesting category is industrial boilers that can be run on alternative energy sources including electricity and that can be switched from one source to another in a relatively short time.

Assuming a time resolution of one hour we can portray the short-run demand by looking at the variation in energy use hour by hour during a day. The use varies over the day, with the lowest energy consumption during the night and a peak at breakfast time and the start of the working day, and again a peak around suppertime on winter weekdays. Figure 1.2 illustrates the power use in Norway for four different days: a summer and a winter weekday and summer and winter Sundays in July and January 2005, respectively. On Sundays the peak demand starts later and on a lower level than weekdays and consumption is somewhat more stable.[6]

[6] The daily load curves for a summer and a winter day (no dates are given) reported in Green and Newbery (1992), Figure 1, p. 935, show the dominant peak to be around suppertime for the winter day and breakfast the summer day. When

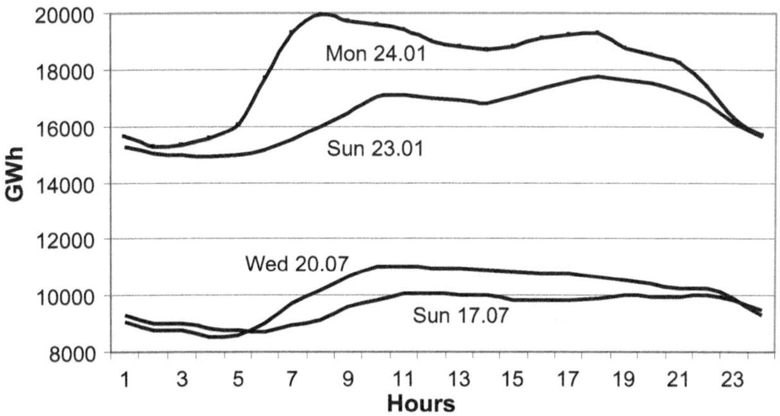

Figure 1.2. Daily load curves. Norway 2005.[7]
Source: Nord Pool

The difference in levels between summer and winter days is considerable and is due mainly to residential heating. It is also more energy consuming to heat water in the winter. The peak in the morning is due to space heating being turned up in wintertime, switching on lights, taking showers, cooking tea, coffee, etc. in dwellings, and then the same for offices (except showering). The afternoon power increase is turning up the room heating again in wintertime, switching on lights, TV, etc. and cooking meals. The difference between a weekday and a Sunday in January is probably mainly due to most offices and light industries being closed on Sundays. The ratio between night time lows and daytime highs are 0.84 and 0.87 for winter and summer weekdays, respectively, and 0.77 for both winter and summer Sundays. The lowest night time use comes later on a Sunday than on a weekday, and the peak is around dinnertime in the winter, but breakfast time in the summer. The summer Sunday curve shows a higher night time use than a weekday from eleven o'clock at night to six o'clock in the morning, and the peak is around noon. There are exciting "electric" Saturday nights and slow starts on Sundays in summer in Norway!

Although the system operator may take daily demand profiles as given in the short run, for economists it is natural to assume that demand is not

---

comparing with Norway the use of natural gas in English households should be borne in mind.

[7] Since the load is for one hour it is measured in GWh in Norwegian statistics although load is a power concept.

totally physically given, i.e., based on "needs," but that demand will also depend on price.

To see the need for power capacity it is common to look at hourly consumption for one year and sort the 8760 hours according to the highest consumption first and then in decreasing order. Such a curve is termed the *load-duration curve*. The hours with the highest energy consumption constitute the *peak load*, and the hours with the lowest consumption show the *base load*. In between we have the *shoulder*. The transmission network and generating stations have power capacity limits that must be able to meet peak load demand. Figure 1.3 illustrates the load-duration curve for Norway in 2005.[8] The load curve is declining rather evenly with no pronounced segments except at the very start and end, so peak, shoulder, and base load periods must be defined on an ad hoc basis. Should only the highest load be termed peak and the lowest load base, or, in view of the variability of these levels over years, some intervals of extreme loads be included? The lowest consumption is 8281 MWh in the hour from six to seven o'clock in the morning July 10. This can be defined as base load. Some heavy industrial users of electricity have continuous operation most of the year and close down only for periodical maintenance. The highest consumption is for mornings and afternoons winter days in the months

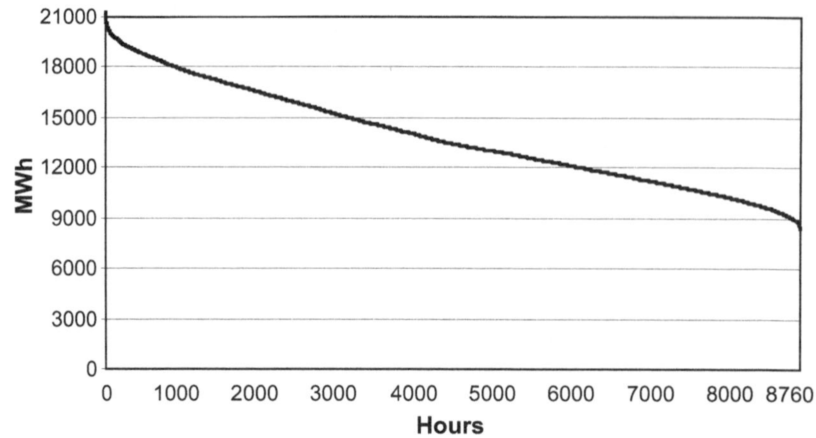

Figure 1.3. The load-duration curve for Norway year 2005.
Source: Nord Pool

[8] In continuous time the load should be measured in MW. The reason for measuring load in energy units, MWh, and not power units (MW) in Figure 1.3 is given in the previous footnote.

December to March. There were 68 hours with a demand above 20000 MWh. The highest demand was for March 2 from eight to nine o'clock in the morning with 21401 MWh. (The highest demand recorded so far is 23054 MWh during a winter morning hour in 2001.) The total capacity is about 25000 MW. The location of hours and dates corresponds to what we saw in Figure 1.2. Peak load is 158% higher than base load within a yearly period.

Hydropower

Electricity generators can use water, fossil fuels, bio fuels, nuclear fuels, wind, and geothermal energy as primary energy sources to run the turbines producing electricity. Hydropower is based on water driving the turbines. The primary energy is provided by gravity and the height the water falls down on to the turbine. Hydropower can be based on unregulated river flows, or dams with limited storage capacity above the natural flow, and on water drawn from reservoirs that may contain up to several years worth of inflow. The total storage capacity in Norway is about 71% of average yearly inflow (excluding minimum storage requirements).

The potential for electricity of one unit of water (a cubic meter) is associated with the height from the dam level to the turbine level. The reservoir level will change somewhat when water is released and thus influence electricity production. Electricity production is also influenced by how water is transported away from the turbine, allowing new water to enter. The turbine is constructed for an optimal flow of water. Lower or higher inflow of water may reduce electricity output per unit of water somewhat. We will return to these issues shortly.

The key economic question in hydropower production is the time pattern of use of water in the reservoir given the production capacity for each time period. With enough storage capacity the water used today can alternatively be used tomorrow. The analysis of hydropower is therefore essentially a *dynamic* one. This is in contrast with a fossil fuel (e.g., coal-fired) generator. Assuming that the market for the primary energy source functions smoothly, running a conventional thermal generator is not a dynamic problem, but is a problem solved period by period, disregarding adjustment cost going from a "cold" state of not producing electricity to a "hot" state producing and back again. In a detailed analysis with fine time resolution the start-up and closing-down costs of thermal units will give rise to dynamic problems, but of a considerably more limited nature than for hydropower.[9]

We are going to use discrete time. This is the case for all practical applications of the type of model we are analysing. From a technical point of view it allows us to use standard mathematics of non-linear programming. The variables are going to be of two types, flow and stock. Stock variables must be dated, e.g., either at the start or at the end of a period. The flow variables will be interpreted as magnitudes related to realisation during a period.

The variables we are going to use are the amount in the reservoir, R_t, inflow of water, w_t, electricity production from regulated hydro, e_t^H, unregulated river, e_t^R, wind power, e_t^W and thermal capacities, e_t^{Th}, respectively. Flow variables in lowercase letters are understood to refer to the period indexed t, while stock variables in capital letters refer to the *end* of the period, i.e., water inflow w_t takes place *during* period t, while the content of the reservoir R_t refers to the water at the end of period t. Release of water, r_t, during period t is converted to electricity (e_t^H) measured in kWh, reflecting the vertical height from the centre of gravity of the dam and to the turbines. The vertical height from the upper level of the dam to the outlet of water from the turbine is called the *gross head* of the reservoir (net head takes into consideration losses due to frictions within tunnels (5%) and the efficiency of turbines (4-5%), electricity generators and systems (2%), tailwater (1-2%), amounting to 12-14% loss of potential energy.

The transformation of water into electricity for a plant with a reservoir can be captured in the simplest way by the production function:

$$e_t^H = f_t(r_t, h_t), f'_{r_t} > 0, f'_{h_t} > 0, \tag{1.1}$$

where r_t is the release of water from the reservoir during time period t and h_t is the gross head. The vertical height of a waterfall is in Norwegian statistics measured from the intake to the turbines and to the release of the tailwater. However, the height from the intake to the level of the dam is also influencing the energy potential of the water. Topology and the constructed wall of the dam give the height so it may be included in the functional form. Then the production function can be given the simple form:

$$e_t^H \le \frac{1}{a} r_t \tag{1.2}$$

[9] If the current price falls below the variable cost of operating a thermal unit it may still pay to keep it running if the price increases again and the loss is less than the adjustment costs; see Chapter 5.

where a is the *fabrication coefficient* (Frisch, 1965), or unit requirement or input coefficient for water; i.e., how many cubic meters (m^3) of water are needed to produce 1 kWh of electricity. If the power station does not have a reservoir, i.e., if it is based on a river flow, then the inflow variable w_t is substituted for the release of water in (1.1) or (1.2).

Neither real capital nor other current inputs like labour and materials are entered in the production function. The role of capital is to provide a capacity to produce electricity; therefore is can be suppressed in an analysis of managing the given capacity. Technology is typically embodied in the capital structure. Turbines represent a quite mature technology and the pace of technical change is now rather slow. The fabrication coefficient will reflect the embodied technology of feeding tunnels and turbines, and the engineering design of optimal water release on to the turbines. We will disregard detailed engineering information about the variation of energy conversion efficiency according to utilisation of a turbine; ranging for 80% for a low utilisation to 95-96% maximally, and then a reduction again if more water is let on to the turbine.

The nature of the costs is important for optimal management of current operations. Given that capacities are present and fixed, only variable costs should influence current operations. However, our specification (1.1) does not show any input other than water, and the water is not bought on a market. Empirical information indicates that traditional variable costs, i.e., costs that vary with the level of output, can be neglected as insignificant. People are employed to overlook the processes and will be there in the same numbers although the output may fluctuate. Maintenance is mainly a function of size of capital structure and not the current output level. (However, wear and tear of turbines depends on the number of start-ups.) We will therefore assume that there are zero current costs. This is a very realistic assumption for hydropower. Water represents the only variable cost in the form of an *opportunity* cost as mentioned above, i.e., the cost today is the benefit obtained by using water tomorrow.

The reduced electricity conversion efficiency due to a reduced height (head) the water falls as the reservoir is drawn down is disregarded. For the Norwegian system, with relatively few river stations and high differences in elevations between dams and turbine stations of most of the dams (the average height is above 200 meters), this is an acceptable simplification at our level of aggregation. In more technically-oriented analyses it may be specified that the coefficient varies with the utilisation of the reservoir (and also with the release of water due to the construction of the turbine giving maximal energy productivity at a certain water flow, as mentioned above). A more detailed analysis taking variable head into consideration for a time period may therefore use an average expression for the fabrication coefficient:

$$a_t = a(r_t, \tilde{R}_t),\ \tilde{R}_t = \frac{R_t + R_{t-1}}{2}, \frac{\partial a}{\partial \tilde{R}_t} < 0, \frac{\partial a}{\partial r_t}\begin{bmatrix} \geq \\ \leq \end{bmatrix} 0,\ t = 1,..,T \tag{1.3}$$

where $\tilde{R}_t$ is the average content of the reservoir during period t. Increasing the release may have either a positive or negative effect on the fabrication coefficient depending on how the release deviates from the optimal design intake of the turbine. An increasing average content of the dam during period t will decrease the fabrication coefficient and increase electricity output at the margin.

In the production-function specification (1.2) we have opened up for waste of released water. However, in the following we assume that the production function holds with equality and then there is a one to one correspondence between water measured in m^3 and water measured in kWh.

The dynamics of water management is based on the filling and emptying of the reservoir[10]:

$$R_t \leq R_{t-1} + w_t - r_t,\ t = 1,..,T \tag{1.4}$$

The amount of water in the reservoir at the end of period t is equal or less than the amount of water at the end of period $t - 1$ (equal to the reservoir content at the start of period t) plus the inflow during period t subtracted the release of water from the reservoir during period t. Evaporation from the reservoir is not accounted for. This is quite reasonable for a northern country like Norway, but may be dealt with in the definition of inflow. In order to save on variables, overflow is not specified as a separate variable. Strict inequality means that there is overflow.

Since maximal head is obtained when the reservoir is full and not influenced by overflow we can in general substitute for water stocks in (1.3) as if equality holds in (1.4):

$$a_t = a(r_t, \tilde{R}_t) = a(r_t, \frac{w_t - r_t}{2} + R_{t-1}) \Rightarrow \frac{\partial a_t}{\partial r_t} = \frac{\partial a}{\partial r_t} - \frac{1}{2}\frac{\partial a}{\partial \tilde{R}_t} \tag{1.5}$$

---

[10] If the stock variable is dated at the *start* of the periods Equation (1.4) will read: $R_{t+1} = R_t + w_t - r_t + s_t$, $t = 1,\ldots,T$, where s_t is the overflow during period t. The stock variable at the *beginning* of period $t+1$ as a function of variables dated t may be a more common way of writing the equation of motion for dynamic systems (Sydsæter et al., 2005).

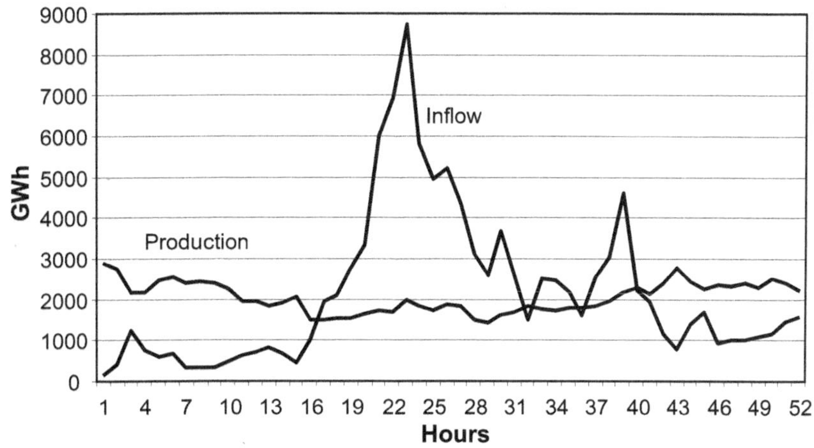

Figure 1.4. Weekly inflow and production of hydropower in Norway 2003.
Source: OED: Fakta 2005

Keeping inflows constant, the change in release is driving both the running operational efficiency and the changing head effect. Since the last derivative on the right-hand side is negative, increasing the release contributes to a lower electricity production at the margin through the head effect.

The annual profile of inflows (w_t) and releases (r_t) in energy units for the Norwegian hydropower system are shown weekly for the year 2003 in Figure 1.4. The water flows are converted to energy units by division with the average fabrication coefficient for the Norwegian hydro system. The inflows are low in the winter weeks, with frost from about the end of October to end of April. In that period production is higher than inflow and this condition lasts until the beginning of August, when all the snow usually has melted in the mountains. Production is greater than inflows in weeks 1 to 16 in Figure 1.4. In the autumn there is rainfall with positive build-up of reservoirs, with weeks 32 and 36 as exceptions in 2003. From week 39 (first week of November) to the end of the year the inflows fall short of the production. The role of the dams is to permit a transfer of water from the late spring, summer, and early autumn weeks to the late autumn and winter weeks. The peak of the snow melting in 2003 was in the beginning of June (week 23). In Norway the snow melting during a few spring and summer weeks fills the reservoirs with about two thirds of the yearly total. The week with the lowest production, week 29 (the last week in July), has 49% of the production of the maximal week 1 (the first week of January). The variation in inflow is much more pronounced, with the lowest inflow in week 1 being 2% of the highest inflow in week 23 (beginning of June).

The production of electricity expressed by (1.2) and the water dynamics of (1.4) are valid for any length of the time period. In studies of optimal management of stored water for electricity production the period concept may be as crude as two aggregate periods of a year (summer and winter seasons based on difference in inflow and/or release profile), and anything from months, weeks, days, and down to hours. A realistic modelling (e.g., the Norwegian total system model; see Haugstad et al., 1990); Gjelsvik et al., 1992); Wangensteen, 2007) may use a week as a period unit and involve a horizon of three to five years.

Environmental concerns

Hydropower is often termed green energy because its production does not generate harmful emissions such as regional pollutants like SO_2 and NO_x or a global pollutant like CO_2. However, although there may be also current environmental problems with hydropower, the main environmental problem is the exploitation of hydropower sites as such. Reservoirs are often artificially created, flooding former natural environments or inhabited areas, although in Norway many reservoirs are based on natural lakes in remote mountain areas. Furthermore, water is drained from lakes and watercourses and transferred through tunnels over large distances, and finally there are the pipelines from the reservoir to the turbines that often are visible, but they may also go inside mountains in tunnels. Thus hydropower systems “consume” the natural environment itself. The waterfalls, lakes and rivers that tourists enjoyed are not there any more. There may also be current environmental problems due to the change in the reservoir level and the amount of water downstream. Changing reservoir levels may create problems for aquatic life, as may also changing levels of release of water downstream, in addition to problems for agriculture in changing the microclimate in the areas of the previously natural rivers and streams.

The conflict between environmental groups and the authorities wanting to exploit waterfalls finally led to a political solution in Norway with compilation of a list of waterfalls that will not be exploited adopted by the parliament in the mid 1980s. The protected waterfalls amount to about 51% of remaining hydro resources to be exploited. The unprotected waterfalls represent an increase in average yearly production of 35%. [Both figures refer to the situation at the start of 2005; see OED, 2005).]

Chapter 2. Water as a Natural Resource

The basic hydropower model

Some studies of hydropower at a high level of aggregation disregard the storage process and specify directly the available water within, e.g., a yearly weather cycle. The assumptions are then that there is no spill of water or binding upper reservoir constraint, and no emptying of the reservoir until the terminal period. The modelling can then be simplified by disregarding the water-accumulation relation (1.4). Another way for this specification to make sense in our framework would be for all the water to be present in the first period. The time profile of inflows should be such that the bulk of inflow comes in one period and then there is a natural seasonal precipitation cycle with little inflow until one year later. The snow, melting during a few spring and summer weeks, fills the reservoirs with about two thirds of the yearly total in Norway. This is illustrated in Figure 1.4 in Chapter 1. The inflow is low in other periods except for autumn rains. However, there are huge variations up to ± 30% from year to year in the pattern of inflow.

In the case of all water being present in the first period, utilisation of water within a horizon can be regarded as a problem of managing a resource of finite amount, just like extraction of non-renewable resources like oil.

As can be seen from Figure 1.4 the validity of the assumption of inflow in only one period depends on the length of the time period. The time periods can be arranged such that inflow occurs in the first period. The basic model is then obtained by assuming that there is inflow only in the first period, and furthermore we assume that the production of electricity is efficient, i.e., we have equality in the production function (1.2). Finally there is unlimited transferability of water to the other periods of the given total amount of water available after the first period. The sum of all releases must then equal the inflow in period 1. Using the production function (1.2) yields:

$$\sum_{t=1}^{T} r_t = w_1 \Rightarrow \sum_{t=1}^{T} a e_t^H = w_1 ,$$
$$\sum_{t=1}^{T} e_t^H = \frac{w_1}{a} = W \tag{2.1}$$

The horizon, T, is assumed to cover a seasonal cycle (one year) from spring to spring. In the first line of equation (2.1) water is measured in m^3, while in the second line of (2.1) water is measured in kWh by using the fabrication coefficient from (1.2) as deflator. Although the variable, W, representing total available inflow, is measured in energy units, kWh, we will still call W water. By assuming no wasting of water as a factor of production in producing electricity, the conversion from water to electricity does not have to be modelled as a separate relationship, but production substituted for the releases as in (2.1).

We will investigate the resource use problem as a standard social planning problem. The energy consumption in each period is evaluated by utility functions, which can be thought of as either valid for a representative consumer or constituting a welfare function. Simplifying further, there is no discounting. The horizon is at any rate usually too short for discounting to be of practical significance (however, Norway has a large proportion of multi-year reservoirs, implying that a rather long horizon, usually three to five years, is warranted). The period utility functions representing the social value of electricity consumption are:

$$U_t(e_t^H) \ , \ U_t'(e_t^H) \geq 0 \ , \ U_t''(e_t^H) < 0 \ , \quad t = 1,..,T \tag{2.2}$$

The utility functions have the standard property of concavity. The marginal utility U_t' measured in monetary units, is defined as the marginal willingness to pay, p_t, i.e., defining the *demand function* (on price form) for electricity:

$$U_t'(e_t^H) \equiv p_t(e_t^H) \tag{2.3}$$

The marginal willingness to pay for electricity is also referred to as the social price (p_t) of electricity or price for short below. We will assume that this demand function has normal properties, e.g., decreasing in quantity corresponding to the assumption about the curvature of the utility function. In light of the brief discussion in Chapter 1 about the sensitivity of demand for electricity to current price, the time period considered should not be too short.

The social optimisation problem can be formulated as follows:

$$\begin{aligned}&\max \sum_{t=1}^{T} U_t(e_t^H)\\&\text{subject to}\\&\sum_{t=1}^{T} e_t^H \leq W, e_t^H \geq 0, t=1,..,T\\&W, T \text{ given}\end{aligned} \tag{2.4}$$

The horizon ends at T and there is no amount of water handed over to period $T + 1$. This assumption may be acceptable if the number of periods T corresponds with almost emptying the reservoir levels due to typical seasonal variation in inflows. (Introducing a lower constraint on water handed over and/or specifying a scrap-value function will be followed up in Chapter 3). The endogenous variables are the electricity production (corresponding uniquely to water use) in each period. To find a solution to the optimisation problem above, we will use a standard nonlinear programming approach (see Sydsæter et al., 1999, 2005).

The Lagrangian function for problem (2.4) is:

$$L = \sum_{t=1}^{T} U_t(e_t^H) - \lambda\left(\sum_{t=1}^{T} e_t^H - W\right), \tag{2.5}$$

where λ is the Lagrangian parameter. Necessary first-order conditions for this problem, where all the variables are non-negative, are:

$$\begin{aligned}&\frac{\partial L}{\partial e_t^H} = U_t'(e_t^H) - \lambda \leq 0 \;(= 0 \text{ for } e_t^H > 0), t=1,..,T\\&\lambda \geq 0 \;(= 0 \text{ for} \sum_{t=1}^{T} e_t^H < W)\end{aligned} \tag{2.6}$$

The endogenous variables are e_t^H, λ ($t = 1,\ldots,T$), $T + 1$ variables in all, and exogenous variables are W, T, two in all. The number of equations is the T first-order conditions in (2.6) and the resource constraint from (2.4), yielding as many endogenous variables as equations. We are conducting a qualitative analysis assuming that a unique solution to problem (2.4) exists. A sufficient condition for a solution to problem (2.4) is that the Lagrangian (2.5) is concave, which is satisfied under our assumptions. Therefore we focus our attention on interpreting the first-order conditions (2.6).

From the Kuhn – Tucker conditions (2.6) we know that the marginal utility of electricity consumption is equal to the shadow price on the resource constraint if we have an interior solution for the energy consumption for period t, i.e., $e_t^H > 0$. The shadow price on the resource constraint is zero if the constraint is not binding. The general interpretation of a shadow price on a constraint is that it shows the change in the objective function of a marginal change of the constraint. In our case the shadow price shows the increase in the sum of utilities over all periods of a marginal increase in stored water, W.

In such a highly stylised model as above it is reasonable to assume that there is positive consumption of electricity in each period and that consumption is not satiated, i.e., that marginal utility is positive in all periods. It then follows that the shadow price on the resource constraint must be positive. The typical conclusion in this basic model with a given amount of resources is that the marginal utility of electricity is constant and equal for all periods:

$$U_t'(e_t^H) = \lambda \text{ for all } t = 1,..,T \tag{2.7}$$

As mentioned above when measuring utility in money, marginal utility may be interpreted as the demand function for electricity on price form. The result of the basic model can then be equivalently stated as the price of electricity being the same for all periods. This is *Hotelling's rule* for the resource price for our model. We do not discount, and by arbitrage of the water asset the social price must be the same for all periods. If prices were different, then, by the assumption of unlimited transferability of water between the periods, transferring water to high-price periods will increase welfare until the prices are equalised in the optimal solution. The shadow price on the water resource constraint measures the increase in the sum of utilities of a marginal increase in the resource, and due to perfect transferability between periods there is only one shadow price.

The typical solution for both periods is illustrated in Figure 2.1 in the case of two periods via a *bathtub* diagram. The two marginal-willingness-to-pay-functions are measured along the left- and right-hand vertical axes for period 1 and period 2, respectively. Total available electrical energy in kWh for the two periods corresponds to the horizontal length of the bathtub. The economic interpretation of the solution to the allocation problem is that electricity should be allocated between the periods in such a way that the shadow price of electricity (i.e., the increase in the objective function of a marginal increase in the given amount of total energy) is equal to the marginal utility of energy in each period, and thus the marginal utilities become equal. In Figure 2.1, if period 1 is summer and period 2 winter, the marginal utility should be equal. Although the marginal utility of energy

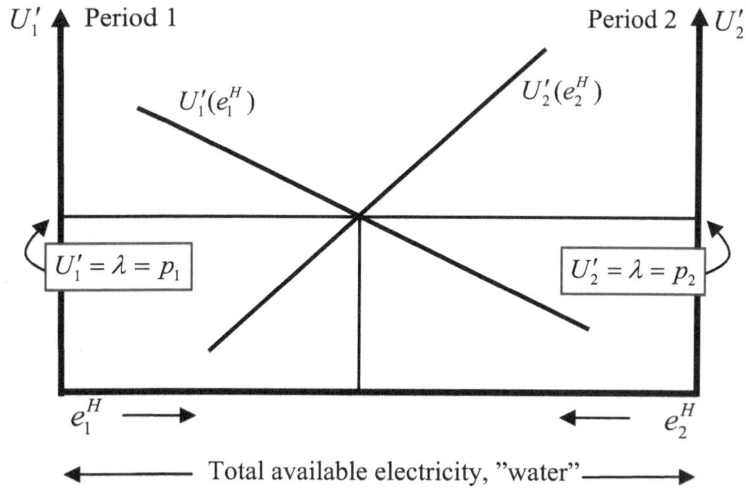

Figure 2.1. Bathtub illustration of optimal allocation of electricity between two periods.

consumption may be higher in winter than in summer for the same level of consumption, marginal utility in the winter should not become greater than in summer in the optimal solution. The consumption in the winter may be substantially higher than in the summer, just as we saw in Figure 1.2 in Chapter 1 for summer and winter days and in Figure 1.4 for weekly periods. The solution for the shadow price is such that all available water is just used up for electricity production.

Water as a non-renewable resource: Hotelling revisited

In the problem (2.4) above water appears as if it is a non-renewable resource with a known initial deposit like oil or minerals since the horizon ends at *T*. The Hotelling rule for a change in the price of a non-renewable resource is usually stated as requiring the resource price to increase with the discount rate. We introduce discounting in our model to show how the familiar form of the Hotelling rule can be derived. Denoting the discount factor β_t we have the following optimisation problem:

$$\max \sum_{t=1}^{T} U_t(e_t^H)\beta_t$$

$$\text{subject to} \qquad (2.8)$$
$$\sum_{t=1}^{T} e_t^H \leq W$$

The discount factor is in discrete time specified as

$$\beta_t = (1+r)^{-(t-1)}, \ t = 1,..,T \ , \qquad (2.9)$$

where r is the rate of discount, assumed to be the same for all periods. The utilities are discounted to period 1, so the discount factor for this period is 1. Notice that the discount rate must correspond to the period length in question, e.g., if a yearly rate is 5%, then if the time period is a week, using the rule for compound interest rate, the weekly discount rate is $r = 0.0009$ and $\beta_2 = 0.999$.

The first-order conditions are straightforward extensions of (2.6):

$$\begin{aligned} &\frac{\partial L}{\partial e_t^H} = U_t'(e_t^H)\beta_t - \lambda \leq 0 \ (= 0 \text{ for } e_t^H > 0), t = 1,..,T \\ &\lambda \geq 0 \ (= 0 \text{ for } \sum_{t=1}^{T} e_t^H < W) \end{aligned} \qquad (2.10)$$

The discounted marginal utilities shall be set equal for all periods and equal to the shadow price on the water resource constraint. The shadow price now measures the change in the *discounted* sum of utilities of a marginal change in the amount of the resource.

The growth rate in marginal utility is found by first using the first-order condition (2.10) for period t and $t+1$ substituting for the discount factor from (2.9):

$$\begin{aligned} &U_t'(e_t^H)\beta_t = U_{t+1}'(e_{t+1}^H)\beta_{t+1} \\ &\Rightarrow U_{t+1}'(e_{t+1}^H) = U_t'(e_t^H)\frac{\beta_t}{\beta_{t+1}} = U_t'(e_t^H)(1+r) \end{aligned} \qquad (2.11)$$

The growth rate in marginal utility from period t to period $t+1$ is then:

$$\frac{U_{t+1}'(e_{t+1}^H) - U_t'(e_t^H)}{U_t'(e_t^H)} = \frac{U_t'(e_t^H)(1+r) - U_t'(e_t^H)}{U_t'(e_t^H)} = r \qquad (2.12)$$

The growth rate is the rate of discount, just as the Hotelling rule tells us about the resource price. Remembering that the marginal utilities by definition (2.3) are interpreted as prices, we have established the Hotelling rule:

$$\frac{p_{t+1}(e_{t+1}^H) - p_t(e_t^H)}{p_t(e_t^H)} = r \tag{2.13}$$

In light of the results of the previous section it should be emphasised that without discounting the fundamental insight of the Hotelling rule for the asset equilibrium, at least for time spans of restricted length, is not really the price growth, but the *level* of the prices. Empirical investigations of resource price development that only check the rate of growth are not so interesting unless the optimal level of prices is checked, too.

An illustration of the consequence of discounting is set out in Figure 2.2 for two periods. The optimal situations without discounting from Figure 2.1 are shown by the dotted lines. The discount factor is one in period 1. In period 2 the discount factor means that the discounted demand curve constitutes a downward vertical shift of the demand curve with the distance

$$U_2'(e_2^H) - U_2'(e_2^H)\beta_2 = U_2'(e_2^H)(1-(1+r)^{-1}) = U_2'(e_2^H)\frac{r}{1+r} \tag{2.14}$$

This curve is shown as the solid curve in Figure 2.2 for period 2. For period 1 the marginal utility and the price are equal to the shadow price on the total water resource. The allocation of electricity in the two periods is determined by the intersection of the demand curve for period 1 and the shifted demand curve for period 2. We see that discounting implies that

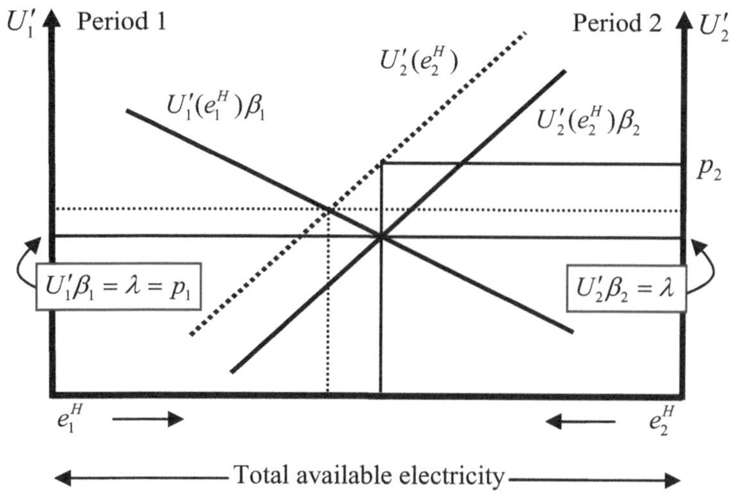

Figure 2.2. Bathtub illustration of the Hotelling rule with discounting. Situation without discounting shown by thin dotted lines.

more is consumed in the first period and less in the second compared with a situation without discounting. The shadow price on the water resource is lower with discounting. This reflects the fact that discounting means more of the resource is preferred to be consumed earlier, and to realise this, prices in earlier periods must be decreased. The price for period 2 is found by going up to the period 2 demand curve. The period 2 price is higher than the period 1 price in accordance with the Hotelling rule.

An interesting economic question is how the endogenous variables change in response to changes in exogenous variables. The consequence of a change in the rate of discount can be found by differentiating the discount factor (2.9) with respect to the rate of discount:

$$\frac{\partial \beta_t}{\partial r} = \frac{\partial (1+r)^{-(t-1)}}{\partial r} = -(t-1)(1+r)^{-t} < 0 \ (t = 2,..,T) \tag{2.15}$$

The reduction in the discount factor (increase in the rate of discount) means that future periods count less in the objective function in the optimisation problem (2.8). The effect is illustrated in Figure 2.3, based on Figure 2.2. The dotted lines represent the situation before an increase in the rate of discount and the solid lines the situation after the increase. The dotted demand curve for period 2 reflects the value of the discount factor before the change and the solid demand curve reflects the value of the discount factor after the change. With less emphasis on the future more will be consumed in the first period. The price then has to go down in the first

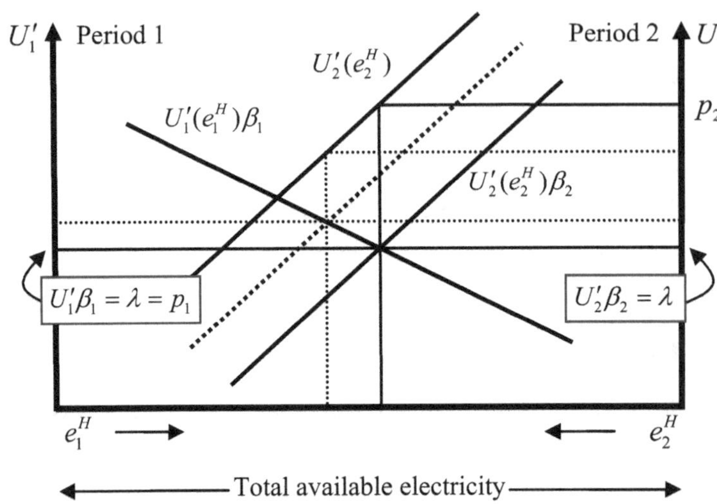

Figure 2.3. An increase in the rate of discount. Situation before change shown by dotted lines.

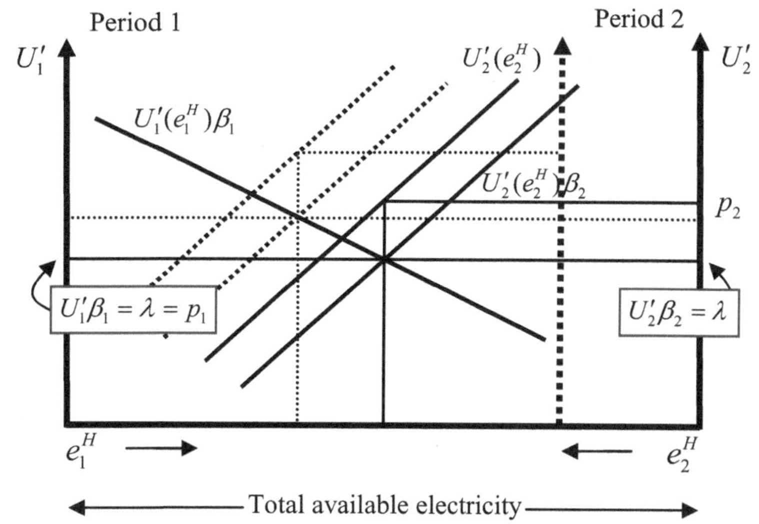

Figure 2.4. An increase in the amount of the resource. Situation before increase shown by thin dotted lines.

period, and the price is increased in the second period to match the decease in the availability of electricity. There is a downward vertical shift in the discounted demand curve for period 2, as is also evident from the expression (2.14) for the distance between the marginal-willingness-to-pay curve and the discounted curve for period 2.

An increase in the availability of water can in the two-period case be illustrated by letting the bathtub wall for period 2 shift outwards to the right in Figure 2.2 as done in Figure 2.4. The dotted curves illustrate the situation in Figure 2.2 before the change in the availability of water, while the two solid demand curves for period 2 represent the situation after the increase in the resource. The shadow price on the resource decreases, as do both period prices. Consumption in each period correspondingly increases.

Several user groups

In the model the water is used to (costless) produce electricity. But we may also consider preferences for water directly by substituting the release r_t for electricity in the utility functions. Water resources are also of interest for activities other than producing electricity. An interesting problem is then how to allocate water if there is competing interest for the water resource. Broad water use groups may be households, industry, agriculture and hydropower. In the case of drinking water for households the interest may lie in the utility of different groups of households, for instance representing different income groups or living within specific locations. As to farmers, industry and hydropower plants, it may be more appropriate to operate with profit functions. However, we will use utility functions for water user groups without being more specific. The G groups are indexed with a superscript g:

$$U_t^g(r_t^g)\,, U_t^{g\prime} \geq 0, U_t^{g\prime\prime} \leq 0,\; g = 1,..,G,\, t = 1,..,T \tag{2.16}$$

The release of water, r_t^g, to each group is drawn from a common reservoir. Utility function can vary over time periods because households' utility of water may vary with outdoor temperature and for agriculture utility may vary with growth season. Industry demand may be more neutral as to time periods.

We will still use the reservoir model (1.4) in Chapter 1, and either assume that all inflows of water occur in the first period or that the upper constraint on the reservoir is never binding and that the reservoir is not emptied until the terminal period. The water constraint can be aggregated into a single one and expressed analogously to (2.1):

$$\sum_{t=1}^{T}\sum_{g=1}^{G} r_t^g \leq W \tag{2.17}$$

Both the total water resource W and the release r_t^g from the reservoir are now measured directly in m^3. The user groups draw water from the same source. The priority given to different user groups is taken care of by specifying a social benefit or welfare function, $B(.)$, constant over time for simplicity, in the utilities of the user groups. This benefit function has the traditional properties from welfare theory, i.e., it is increasing at a decreasing rate in all the utilities. The social planning problem can then be formulated as:

$$\begin{aligned}
&\max \sum_{t=1}^{T} B(U_t^1(r_t^1),..,U_t^n(r_t^G)) \\
&\text{subject to} \\
&\sum_{t=1}^{T}\sum_{g=1}^{G} r_t^g \leq W \\
&r_t^g \geq 0, g = 1,..,G, t = 1,..,T \\
&T, W \text{ given}
\end{aligned} \tag{2.18}$$

It is straightforward to introduce discounting in the model using discount factors such as β_t in the previous section.

The Lagrangian is:

$$L = \sum_{t=1}^{T} B(U_t^1(r_t^1),..,U_t^G(r_t^G)) - \lambda(\sum_{t=1}^{T}\sum_{g=1}^{G} r_t^g - W) \tag{2.19}$$

The necessary first-order conditions are:

$$\begin{aligned} &\frac{\partial L}{\partial r_t^g} = B_g{}'U_t^{g\prime}(r_t^g) - \lambda \le 0 \; (= 0 \text{ for } r_t^g > 0), t = 1,..,T, g = 1,..,G \\ &\lambda \ge 0 \; (= 0 \text{ for} \sum_{t=1}^{T}\sum_{g=1}^{G} r_t^g < W) \end{aligned} \tag{2.20}$$

The shadow price, λ, on the water constraint may now properly be termed the *water value* since it measures the change in the objective function of a marginal increase in the amount of water measured in m^3. Assuming that water is to be consumed for each group in each period we have that the discounted socially weighted marginal utilities of water consumption should all be equal between different user groups and equal over time,[1] and equal to the water value. The water value is the crucial equilibrating variable telling us that the socially weighted value of the marginal utility of drinking water should be set equal to the socially weighted value of marginal utility of irrigation water, equal to the socially weighted marginal utility of industry consumption and equal to the socially weighted marginal utility of hydropower water use.

If the distributional objective expressed by the benefit function is dropped, e.g., by specifying the benefit function as a pure summation of utilities, and in addition assuming that utilities are measured in money, then a total demand function (on price form) can be formed by adding (horizontally) the individual demands. Each group's marginal willingness to pay is now measured in the same unit, money:

$$\sum_{g=1}^{G} U_t^{g\prime}(r_t^g) = \sum_{g=1}^{G} D_t^g(r_t^g) = D_t(r_t) = p_t, \sum_{g=1}^{G} r_t^g = r_t \tag{2.21}$$

An optimal allocation of water between groups for a time period can be illustrated as in Figure 2.5, specifying three groups. The group demand curves derived from the marginal utilities measured in money are drawn as straight lines sloping downwards starting at finite levels at zero consump-

[1] If a discount factor is used, then the socially weighted marginal utilities will change correspondingly over time.

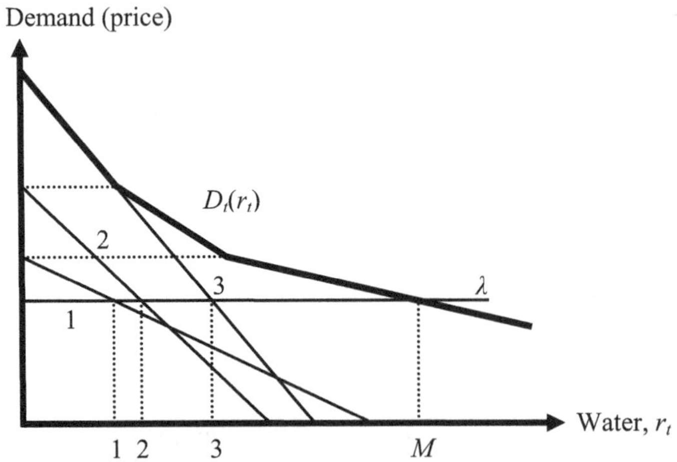

Figure 2.5. Aggregation of individual demand curves. Equilibrium water shares for period t.

tion of water. The individual group allocations are found by the intersections of the demand curves with the common horizontal shadow price line of the water resource. The levels are indicated by 1, 2, and 3 on the horizontal axis. If the shadow price is higher than the choke price, then no water is allocated to this group. The aggregated total demand curve is $D_t(r_t)$ and the total consumption is indicated by the point M.

As to the time allocation problem we could use the bathtub construction for two periods and extend Figure 2.5 to a figure like 2.1. The point of intersection of the aggregate demand curves will coincide with the value of the horizontal line for the shadow value of water. The social prices will be equal for each group for all time periods. The quantities allocated to the groups may vary with the time period, but the social price remains the same. (If discounting is introduced we get the same change in focus to discounted prices being equal as in the previous section.)

The allocation over time is illustrated in a bathtub diagram in Figure 2.6 for two periods. The allocation between the two periods is given by the intersection of the total demand curves and shown by the point M on the horizontal total water axis. The equilibrium price is given by λ and is equal both across user groups and periods. The three groups get the allocation of water in period 1 indicated by the vertical dotted lines marked 1, 2, and 3. The demand structure, keeping roughly the order of period 1, is such that group 1 now does not get any water in period 2. The willingness to pay is not high enough. This may be the case of irrigation water in the rainy sea-

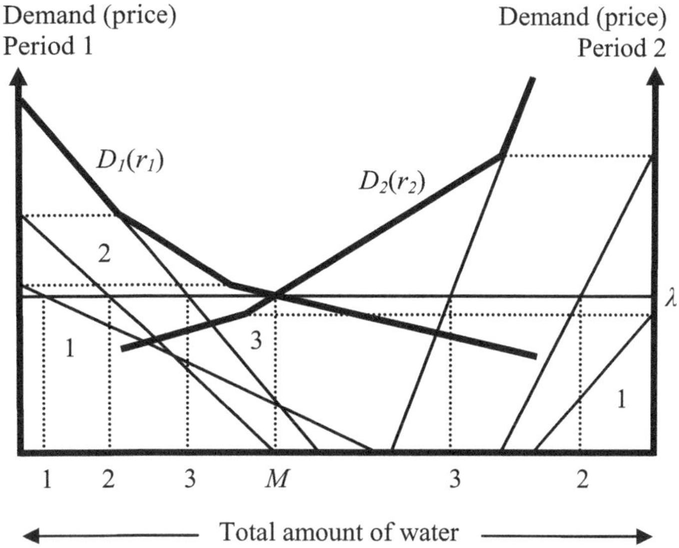

Figure 2.6. Allocation across groups and over time.

son. Groups 2 and 3 get the allocations indicated by thc vertical dotted lines for period 2. Notice that the price is the same even if one period is a drought period and the other is a rainy season. Without uncertainty the water that is collected during the first period is always shared in such a way that the price is the same over time.

Chapter 3. Hydropower with Constraints

The variation of prices

The analysis in Chapter 2 concluded that the price should be the same for all time periods. However, even a superficial knowledge of electricity markets with a significant presence of hydropower tells us that electricity prices vary over seasons and even days. The hourly prices for the four winter - summer days, used in Figure 1.2 in Chapter 1 to show the electricity demand, are shown in Figure 3.1.[1] The price levels of the summer days are, somewhat surprisingly, higher than for the winter days with the exception of a couple of night hours when the winter Sunday prices dip below the

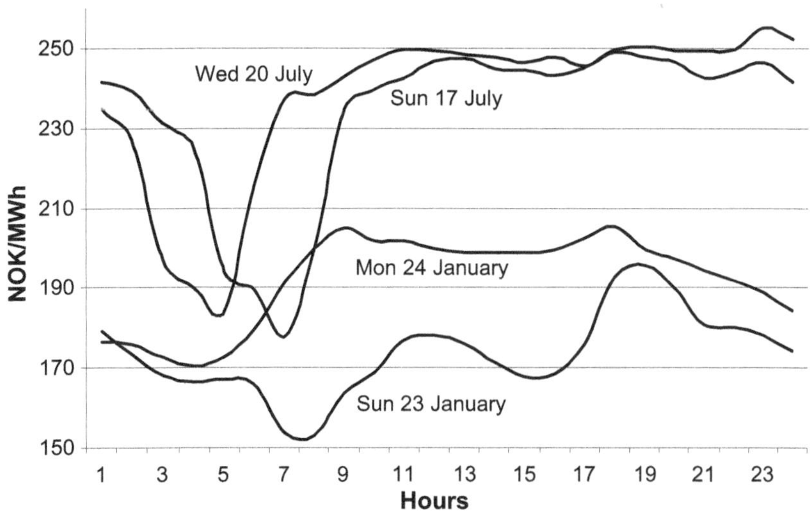

Figure 3.1. Hourly price variation for four days in Norway 2005.
Source: Nord Pool

[1] The Norwegian currency is denoted NOK, and the exchange rate was about 8.15 NOK for one Euro in April 2007.

summer weekday price. [The explanation for the unusual price relationship between the chosen summer and winter days turned out to be exceptionally high inflows in December 2004 and January 2005 (seen in Figure 3.10) leading to exceptionally low prices in January 2005.] All days show lower prices during night hours. This difference is especially pronounced for the two summer days. These days have almost the same price curves, while the winter Sunday and weekday differ considerably more.

To get an impression of the variability of prices over a year the hourly prices are sorted in decreasing order in Figure 3.2 to make a price-duration curve. There are two distinct turning points or knuckles of the curve. The highest price of 725 NOK/MWh (Euro 89) was March 3 eight o'clock at night. Then the price falls to about 290 NOK/MWh (Euro 36) at the first knuckle point, or with 60%. Most of the high prices are for morning hours between seven and ten o'clock and also some afternoon hours. There are 251 hours in the interval 290-725 NOK/MWh (Euro 36-89), or 2.9% of the total 8760 hours. In between the knuckle points the price falls with 41%. The price range 170-56 NOK/MWh (Euro 21-7) covers the steep right-hand part past the second right-hand knuckle. For a few hours the prices are quite lower than the median price 236 NOK/MWh (Euro 29). This part encompasses 266 hours, or 3.0% of the total hours. The lowest price is for November 15 at three o'clock at night. The typical hour for the majority of the low price range is, in fact, during the night.

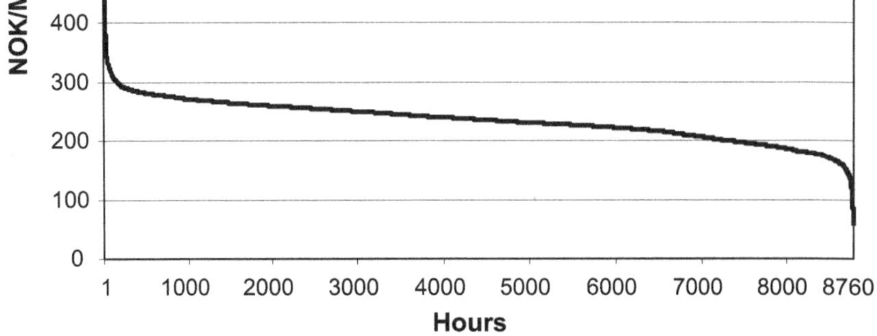

Figure 3.2. Price-duration curve Norway 2005.
Source: Nord Pool

In the perspective of these data we should come up with mechanisms that generate considerable price variations if our model is to be of help to understand actual electricity markets.

Constraints in hydropower modelling

In Chapter 2 any constraints on the reservoirs were suppressed and only a limit on the total available water was used. However, there are many constraints on how to operate a reservoir and a hydropower plant. The relevance of the restrictions will vary somewhat with the length of period chosen for the model, from the more aggregate of two periods within a year to hourly resolution. The relevance of the constraints also depends on whether single or multiple plant models are adopted. The main types of constraints are shown in Table 3.1.

A fundamental constraint is that a maximal amount of water can be stored. This constraint is valid for any level of time resolution, but especially important to include within a longer time horizon. This constraint will have a crucial importance for how the dam can be operated. There is a maximal physical upper limit, but due to, e.g., environmental concerns the limit may vary with period and be below the absolute physical limit for some periods.

Environmental concerns are even more relevant for the lower limit and may impose constraints on how much the dam can be emptied. Empty

Table 3.1. Constraints in the hydropower model.

Variable	Constraint type and variable	Expression
R_t : reservoir at end of t	Max reservoir: $\overline{R}_t$	$R_t \leq \overline{R}_t$
	Environmental concerns, min reservoir: $\underline{R}_t$	$R_t \geq \underline{R}_t$
e_t^H : hydropower during t	Max power capacity: $\overline{e}^H$	$e_t^H \leq \overline{e}^H$
	Max transmission capacity: $\overline{e}_t^H$	$e_t^H \leq \overline{e}_t^H$
r_t : release of water during t	Water flows, environment: $\underline{r}_t = \min$, $\overline{r}_t = \max$	$\underline{r}_t \leq r_t \leq \overline{r}_t$
	Environment: ramping up: r_t^u	$r_t - r_{t-1} \leq r_t^u$
	ramping down: r_t^d	$r_{t-1} - r_t \leq r_t^d$

dams create eyesores in the landscape, and can create bad smells from rotting organic material along the exposed shores. Fish may have problems surviving or spawning at both too low and too high water levels. The environmental lower constraint may depend on the time period, because the environmental problems may vary with season. In Norway, where the dams are covered by ice in the winter season, the lower level may be less than in the summer.

The capacity of a power station may be constrained by the installed turbines or the diameter of the pipe from the reservoir to the turbines. Such a constraint has no subscript for time period. The power concept will follow the period definition. For example, if the period length is 1 hour the power constraint is measured in kWh, by using the maximal kW rating for 1 hour. Using only energy as our variable the power constraint is the same as a production constraint. When ramping up in period t we have $r_t > r_{t-1}$ and when ramping down we have $r_{t-1} > r_t$.

In aggregated analyses it is common not to specify the transmission system. But a constraint on transmission can be represented the same way as for power capacity constraint, except that a time index may be used on the constraint to indicate that transmission capacity within some limits is an endogenous variable governed by physical laws of electrical flows in a multilink grid system between input and output nodes. The loss may also vary with temperature: resistance is higher in hot weather than in cold weather. However, this effect is rather insignificant. The lion's share of loss variation is due to variation in the flow through the lines. We will return to the specification of a network in Chapter 7.

There may be environmental concerns about the size of the release, r_t, from a reservoir. If the release occurs into a river system there may be concerns both about the lower and the higher amount of water that should be released due to impacts on the environment downstream. Impacts on fishing and recreational activity and pressure from tourism may be relevant. Erosion of riverbanks and temperature change for agricultural activity nearby may also count. Then there is concern about navigation and flood control.

All these effects may also be present when releases change, so upper constraints may be introduced both on ramping up, r_t^u, and ramping down, r_t^d. These constraints are most relevant for shorter time periods.

The constraints introduced for environmental reasons may reduce the amount of current environmental problems to a minimum, or below a level where net benefits of further constraints are negative according to a majority view. We will therefore not treat environmental concerns explicitly when studying the hydropower management problem. As mentioned in

Chapter 1, the most severe environmental damages arises constructing the physical hydropower system, and not in the operational phase.

Optimal management with reservoir constraint

In order to study optimal management of the hydro system, an objective function has to be specified. In the older literature on hydropower referred to in Chapter 1 and in engineering literature (Wood and Wollenberg, 1984) the social objective function is often expressed as minimising the total costs of supplying a given amount of electricity within a horizon. In economics a standard objective function in empirical studies is to maximise consumer plus producer surplus with the consumed (equal to the produced) quantities as *endogenous* variables. The consumption side is conveniently summarised by using demand functions[2] [defined in (2.3) in Chapter 2 on inverse form] and the supply side by using variable cost functions. This is a partial equilibrium approach because no interaction with the rest of the economy is modelled. In the case of hydropower with zero operating costs the social surplus is simplified to the area under the consumer demand function (since the consumers' expenditures are identical to the producers' profit):

$$\text{Objective function} : \sum_{t=1}^{T} \int_{z=0}^{e_t^H} p_t(z)\,dz \tag{3.1}$$

We assume that there are no external costs involved in producing or consuming the hydropower. It is assumed that costs that do not depend on the current output level, but can be avoided if the plant is shut down, do not lead to the plant being shut down by the social planner. Such cost terms can therefore be disregarded in the objective function since the optimal solution for running the plant is independent of these cost terms. The use of a demand function relating the period consumption to the same period price is subject to the qualifications mentioned in Chapter 1. A technical assumption needed on the demand functions is that there is a finite choke price yielding zero demand. Otherwise demand is assumed to decrease in price in the standard way in economics. These assumptions are all standard when employing the consumer-surplus concept. Discounting is not introduced since the horizon is usually so short that the effect will be

[2] Notice from Chapter 2 that the demand functions may also be interpreted as representing utility or preference functions.

negligible as pointed out in Chapter 2, but will be straightforward to include, as shown there.

Assuming no waste of water in the production of electricity, the reservoir dynamics is:

$$R_t \leq R_{t-1} + w_t - r_t = R_{t-1} + w_t - ae_t^H \Rightarrow$$
$$\frac{R_t}{a} \leq \frac{R_{t-1}}{a} + \frac{w_t}{a} - e_t^H \tag{3.2}$$

In the last line of (3.2) all the water variables measured originally in cubic meters (m^3) of water are converted to energy units, kWh, by dividing through with the fabrication coefficient, *a*. It will be convenient to express all units in kWh in the rest of the book. However, for notational convenience we will drop explicitly showing the conversion from water units to energy units by suppressing the fabrication coefficient *a*, and still refer to the variables originally measured in water units as "water."

The social planning problem can then be expressed in the following way:

$$\max \sum_{t=1}^{T} \int_{z=0}^{e_t^H} p_t(z)dz$$
subject to
$$R_t \leq R_{t-1} + w_t - e_t^H \tag{3.3}$$
$$R_t \leq \bar{R}$$
$$R_t, e_t^H \geq 0\ ,\ t = 1,..,T$$
$$T, w_t, R_o, \bar{R} \text{ given, } R_T \text{ free}$$

In order to simplify, the reservoir limits are assumed to be independent of period, and the lower level is normalised to zero (i.e., the upper level used in (3.3) is the physical upper level subtracted the lower level; $\bar{R} = \bar{R}_t - \underline{R}_t$). If the lower limit is explicitly modelled then the shadow price on this constraint will tell us the benefit of making the constraint less severe. This information may be useful if there is any discussion or doubt as to the chosen minimum level.

We disregard for the time being all other constraints in Table 3.1. An important consequence is that there is full manoeuvrability of the system in the sense that a reservoir can be emptied within a period. No scrap-value function for water in the reservoir or minimum level in the last period is introduced so far, so the amount at the end of period *T* is free.

The optimisation problem (3.3) is a discrete time dynamic programming problem, and special solution procedures have been developed for this class of problems (Bellman, 1957; Sydsæter et al., 2005). However, because of the special structure of the problem we shall treat it as a standard nonlinear programming problem and use the Kuhn – Tucker conditions for discussing qualitative characterisations of the optimal solution.

The Lagrangian function for problem (3.3) is:

$$\begin{aligned} L = {} & \sum_{t=1}^{T} \int_{z=0}^{e_t^H} p_t(z)\,dz \\ & -\sum_{t=1}^{T} \lambda_t (R_t - R_{t-1} - w_t + e_t^H) \\ & -\sum_{t=1}^{T} \gamma_t (R_t - \bar{R}) \end{aligned} \tag{3.4}$$

Endogenous variables are e_t^H, R_t, λ_t, γ_t ($t = 1,\ldots,T$), and there are $4T$ variables in all. Necessary first-order conditions are:

$$\begin{aligned} & \frac{\partial L}{\partial e_t^H} = p_t(e_t^H) - \lambda_t \le 0 \;(= 0 \text{ for } e_t^H > 0) \\ & \frac{\partial L}{\partial R_t} = -\lambda_t + \lambda_{t+1} - \gamma_t \le 0 \;(= 0 \text{ for } R_t > 0) \\ & \lambda_t \ge 0 (= 0 \text{ for } R_t < R_{t-1} + w_t - e_t^H) \\ & \gamma_t \ge 0 (= 0 \text{ for } R_t < \bar{R}), \qquad t = 1,..,T \end{aligned} \tag{3.5}$$

The number of equations is the $2T$ first-order conditions in (3.5) and $2T$ reservoir constraints from (3.4), so there are as many equations as endogenous variables. As in Chapter 2 we will just assume that the first-order conditions are valid for the optimal solution without going deeper into the mathematics.

Now, our general objective is that the model should tell us something qualitatively about optimal production and consumption of electricity that has real-world interest. We will then limit the number of possible optimal solutions by making reasonable assumptions. One such assumption is that positive production is required in all periods, yielding the conditions:

$$p_t(e_t^H) = \lambda_t \;, \; t = 1,..,T \tag{3.6}$$

The shadow price λ_t of the stored water may be termed the *water value*[3]. It shows in general the change in the value of the objective function, evaluated at an optimal solution, of a marginal change in the constraint. In our case the water value in period t shows the value in terms of an increase in gross consumer surplus of a marginal increase either in the transfer of water from period $t - 1$ or an increase in the inflow in period t. In the engineering literature the expression *system lambda* is used for the marginal generation cost of the electricity system. The water value λ_t as an opportunity cost is just this system lambda.

In the optimal solution with positive production in period t the water value is equal to the social price. Note that by assuming (3.6) we have not ruled out the possibility that the water value is zero. The water value for period t expresses the value of using water in the next period $t + 1$ through the second equation in (3.5). This is the essential dynamic equation for the system. There are only two successive periods involved in the equation of motion. This means that a sequence of two-period diagrams may capture the main features of the general solution. We will use the development of shadow prices to give insights into the qualitative characteristics of an optimal solution.

Introducing terminal conditions

Recognising that "life continues" after the horizon T it is logical to put a terminal condition on the reservoir level for period T. This can be done by introducing a new constraint imposing a minimum level, R^T, or by introducing a scrap value term in the objective function. The constraint added to the constraints in (3.3) is:

$$R_T \geq R^T \tag{3.7}$$

The objective function with a scrap value function becomes:

$$\sum_{t=1}^{T} \int_{z=0}^{e_t^H} p_t(z)\,dz + S(R_T) \tag{3.8}$$

The form of the scrap-value function may be one of a constant marginal value, S', or it may be a concave function with an extreme value on the interior of the interval $[0, \bar{R}\,]$.

---

[3] But remember that in our simplified model water is measured in energy units, kWh. We should really measure water in m^3 to use the expression. This can easily be done by multiplying through with the fabrication coefficient a.

It seems reasonable to assume that the minimum level R^T lies somewhere between zero and the upper reservoir constraint. The first-order conditions involving the constraint (3.7) become

$$\frac{\partial L}{\partial R_T} = -\lambda_T + \omega - \gamma_T \leq 0 \ (= 0 \text{ for } R_T > 0)$$
$$\omega \geq 0 (= 0 \text{ for } R_T > R^T) \tag{3.9a}$$

where ω is the shadow price on the terminal constraint in (3.7) (having the term $-\omega(- R_T + R^T)$ in the Lagrangian). Using the scrap value function instead yields the following condition for the terminal period T, replacing the one stated in (3.5):

$$\frac{\partial L}{\partial R_T} = S'(R_T) - \lambda_T - \gamma_T \leq 0 \ (= 0 \text{ for } R_T > 0) \tag{3.9b}$$

In the case of a minimum level of the reservoir as a constraint in the last period we have that the first condition in (3.9a) holds with equality, and furthermore that the shadow price on the upper reservoir constraint is zero. Leaving more to the future than the minimum reservoir R^T implies a zero value of the shadow price ω, but this can be optimal only if the price in the terminal period becomes zero according to the condition (3.6). Then demand for electricity must be satiated in the terminal period, but we have ruled out this possibility above. Therefore the terminal condition is binding and we assume in the regular case that the shadow price is positive, yielding a positive terminal water value.

In the case of using the scrap-value function the regular case will be that the reservoir level is between zero and the maximal reservoir level, implying that the shadow price on the upper reservoir constraint is zero, yielding equality between the terminal water value and the marginal evaluation for future use of the terminal reservoir level according to (3.9b).

Introducing the minimum level R^T will influence the magnitude of the water value of the terminal period. Instead of adding R_{T-1} to the inflow w_T and then consuming the whole amount in period T, $(R_{T-1} - R^T)$ is now added to the inflow. The range of possible values for the terminal water value is shifted upwards since the maximal production is reduced by the amount set aside for the period after the terminal one. The minimum value of the reservoir handed to the terminal period may now have to be positive in order to fulfil the terminal constraint. We must have min $R_{T-1} \geq R^T - w_T$.

Using the scrap-value function water at the disposal for consumption in the terminal period is $w_T + R_{T-1} - R_T^*$, where R_T^* is the optimal amount left for future use. We get the same type of upward shift of the possible values of the terminal water value as for the case of a minimum level condition.

As we have seen above, introducing a positive minimum terminal value of the reservoir level or a scrap-value function does not change the story about the formation of optimal social prices in principle. Therefore, for ease, we will not use such specifications in this book.

The bathtub diagram for two periods

The conditions (3.5) tell us that there are two events that are crucial for the development of prices and shadow prices: the reservoir running empty and the reservoir running full. Focussing just on two periods can bring this out. The bathtub diagram used in Chapter 2 can now be extended to include a reservoir limit. In the two-period case, assuming that zero spilling is optimal, adding together the two water-storage equations in (3.3) we have

$$e_1^H + e_2^H = R_o + w_1 + w_2 \tag{3.10}$$

The maximal electricity produced is equal to the available water from period $t = 0$ and the inflows in periods 1 and 2. The solution for two periods can be illustrated in a bathtub diagram, Figure 3.3, extending Figure 2.1 in

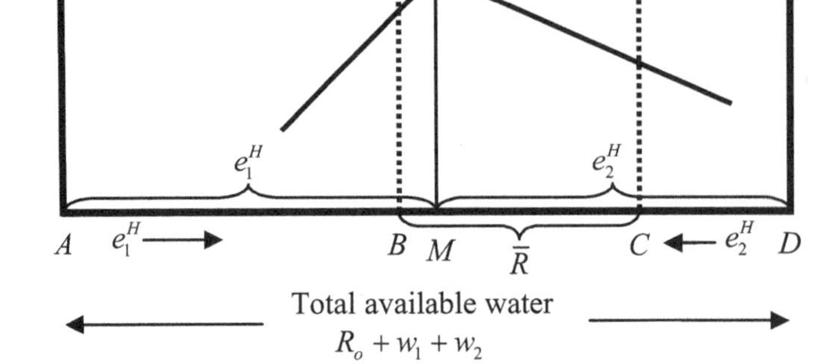

Figure 3.3. Two-period bathtub diagram with a non-binding reservoir constraint.

Chapter 2, showing the total available water as the floor of the bathtub, and the demand curves anchored on each wall. The maximal storage is now introduced. Inflow plus the initial water R_o in period 1 is *AC*, and inflow in period 2 is *CD*. The maximal storage is *BC*. The storage is measured from *C* toward the axis for period 1 because the decision of how much water to transfer to period 2 is made in period 1. The intersection of the demand curves determines the common price for the two periods, equal to the common shadow price on stored water, in accordance with the first-order conditions. The point *M* on the bathtub floor shows the distribution of electricity production on the two periods. The optimal transfer illustrates the case when the reservoir limit is not reached, but there is scarcity in period 2 since all available water, *MC* + *CD*, in that period is used up. Therefore the amount *AM* is consumed in period 1 and *MC* is saved and transferred to period 2. The total amount available for both periods is used up and gives rise to a positive price for both periods, assuming no satiation of demand. The amount consumed in period 1 leaves less than the maximal possible amount to period 2. The intersection of the demand curves takes place within the vertical lines from *B* and *C*, indicating the maximal storable amount. Since water consumed in period 1 is at the expense of potential consumption in period 2 the water values become the same and equal to the price for both periods. We have from (3.5) that $\lambda_1 = \lambda_2$ since $\gamma_1 = 0$ because $R_1 < \overline{R}$, and then from (3.6) we have $p_1 = \lambda_1, p_2 = \lambda_2$.

Expanding the availability of water marginally by expanding one of the inflows will create a value equal to the shadow price on the corresponding reservoir constraint.

The demand curves may also intersect to the left of the broken vertical reservoir capacity line from *B* as illustrated in Figure 3.4. The optimal allocation is now to store the maximal amount *BC* in period 1 because the water value is higher in the second period, and consume what cannot be stored, *AB*, in period 1. Due to the assumption of non-satiation of demand it cannot be optimal with any spill in period 1. From the first-order conditions (3.5) water value and hence the price is zero when having spill. The water value is now higher in the second period. In the second period the reservoir, containing *BC* from the first period and an inflow of *CD* coming in the period, is emptied. We go from a period of threat of overflow to a period with scarcity. Using (3.5) for $R_1 > 0$ we have that $\lambda_1 = \lambda_2 - \gamma_1$. The shadow price on the reservoir constraint, γ_1, is the difference between the water values as indicated in the figure. If the reservoir could be marginally expanded the extra economic value created is the difference between the period prices. (The shadow price on the constraint is the change in the objective function when $\overline{R}$ is marginally changed.)

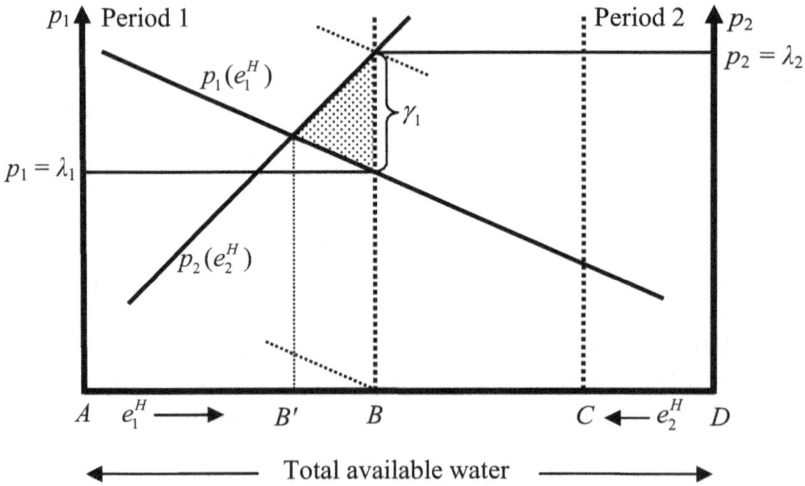

Figure 3.4. Social optimum with reservoir constraint binding.

Notice that the water allocation will be the same for a wide range of period 1 demand curves keeping the same period 2 curve, or vice versa. The period 1 curve can be shifted down to passing through B and shifted up to passing through the level for the period 2 water value, as indicated by the dotted lines as alternative demand curves. The price difference between the periods may correspondingly vary considerably. A binding reservoir constraint implies that the value of the objective function becomes smaller. Using the unconstrained solution as a benchmark, indicated by the vertical dotted line from B' to the intersection of the demand curves, the marked triangle is the reduction in total consumer plus producer surplus due to the limited size of the reservoir.

The bathtub diagram may be used for just two periods as in Figures 3.3 and 3.4, but it may also be used within a multiperiod analysis for two consecutive periods. The two-period nature of the dynamics of the system makes it possible to illustrate a sequence of optimal solutions using two-period bathtub diagrams. Connecting figures like Figures 3.3 and 3.4, we must remember that the inflow AC in the first period now also contains what is stored in the period preceding the one we are studying. In the second period we will now see what is left for the next period.

The generation of price changes

We return to the multiperiod problem for a comprehensive investigation of possible developments of the social price over time. Let us start by investigating the consequences of introducing the upper constraint on the reservoir using model (3.3). According to Bellman's principle for solving dynamic programming problems with discrete time, we start searching for the optimal solution by solving the optimisation problem for the last period and then work our way successively backwards toward the first period.

The terminal period

Although our problem (3.3) is not set up in the standard way for a dynamic programming problem, the recursive structure of the first-order condition for the shadow prices in (3.5) implies that we can solve for the structure of prices and shadow prices by starting with the last period and then work our way backwards. The optimality conditions, using assumption (3.6) for the end period T, are:

$$\begin{aligned} &p_T(e_T^H)=\lambda_T\,(e_T^H>0)\\ &-\lambda_T-\gamma_T\le 0\;(=0 \text{ for } R_T>0) \end{aligned} \tag{3.11}$$

Our horizon ends at T, so the water value for the period $T+1$ does not exist (i.e., is set to zero). For period T we have two possibilities as to the utilisation of the water in the reservoir: either it is emptied, $R_T = 0$, or some water is remaining, $R_T > 0$. Since the water has no value from $T + 1$ on, the latter situation can be optimal only if the marginal utility of electricity becomes zero before the bottom of the reservoir is reached. We will adopt the alternative that the marginal utilities of electricity remains positive to the last drop even if a maximal storage of water is transferred to the terminal period:

$$p_T(w_T+\max R_{T-1})=p_T(w_T+\overline{R})>0 \tag{3.12}$$

This means that we will have a situation of *scarcity* in the last period T with $p_T(e_T^H) = \lambda_T > 0$. Scarcity in an economic sense means that there is a positive willingness to pay for one more unit at the margin (i.e., a small decrease in price would have induced more consumption if more of the good was available). In our situation we also get physical scarcity in the sense that all available water is used up. Scarcity gives economic value to the water in the last period. Since we cannot have a situation of physical scarcity at the same time as the upper limit on the reservoir is reached, the shadow price γ_T on the upper constraint is zero [follows from the last complementary slackness condition in (3.5)]. The second relation in (3.11) then implies $\lambda_T \geq 0$. This does not give us any new information as to the water value in period T (the shadow price may be zero although the expression in the water storage constraint is zero, as is our situation in period T), but by our assumption (3.12) of no satiation in period T the value is positive.

Moving backwards to period $T-1$ the shadow-price equation from (3.5) reads

$$-\lambda_{T-1} + \lambda_T - \gamma_{T-1} \leq 0 \; (= 0 \text{ for } R_{T-1} > 0) \tag{3.13}$$

If we, quite reasonably in a multiperiod setting, disregard the possibility that a full reservoir will be handed over to the terminal period, then γ_{T-1} will be zero. (In a two-period model it may be more probable that a full reservoir is handed over to period 2, as shown in Figure 3.4). If we assume that the reservoir will not be emptied in period $T-1$ then the equation holds with equality, and we have that the shadow price on water in period $T-1$ will be equal to the shadow price in the terminal period T.

The situation of scarcity in one period (period 2) is already illustrated in Figure 3.3. Relabelling period 1 and 2 period T and $T-1$ there is scarcity in period T. Since the reservoir level in period $T-1$ is by assumption at a level between zero and the upper limit, the price and the water values will be the same for period T and $T-1$. Scarcity in period T sets the price for both periods. The water available for period $T-1$, AC, is now made up of the reservoir inherited from period $T-2$, R_{T-2}, and the inflow in period $T-1$. In Figure 3.3, $MC = R_{T-1}$ is transferred to the terminal period T, where MD, consisting of the transfer and the inflow in period T, is consumed.

Neither overflow nor scarcity

Moving backwards in time we will assume that after period $T-1$ we have periods with neither threat of overflow nor emptying of reservoirs. From the necessary conditions (3.5) we then know that the terminal period price p_T will prevail for all these periods. The way such periods can be illustrated is shown in Figure 3.5. AC is made up of inflows in period u plus what is remaining in the reservoir from period $u-1$. CD is the inflow in period $u+1$ and BC is the reservoir capacity. Using sufficiently fine time resolution, the storage capacity may be far greater than the consumption for two consecutive periods. The yearly storage capacity of the Norwegian hydro system of two thirds of average inflow means that production of

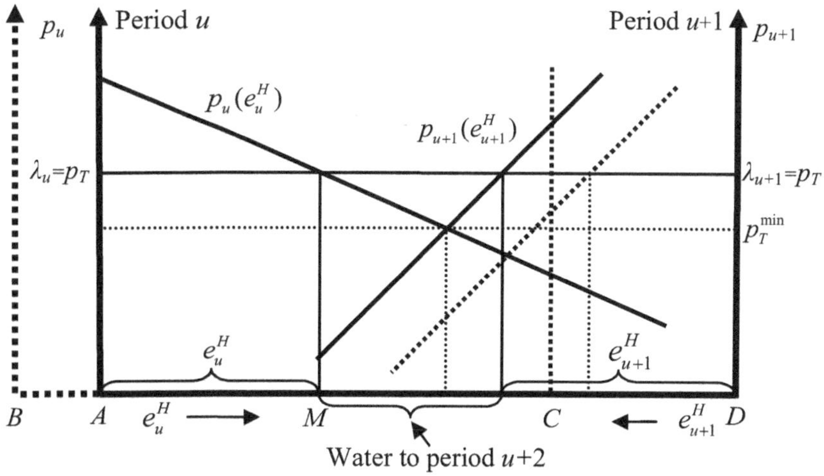

Figure 3.5. Neither threat of overflow nor scarcity.

electricity in, e.g., two consecutive average weeks may be much less in each period than the reservoir capacity. Therefore it will be many periods in which the capacity indicated by *BC* in Figure 3.5 will have *B* to the left of the bathtub wall. This situation implies that it is impossible to run into a period with overflow or threat of overflow. With the price level p_T given from the future this will be the price both in period u and $u + 1$. The amount of water consumed in period u is *AM* and found by the intersection of the period u demand curve and the horizontal price line p_T. The amount indicated by *MC* will be saved in period u for use in period $u + 1$. In period $u + 1$ the inflow *CD* is used up and also an additional amount, as found by the intersection of the demand curve for period $u + 1$ and the price line p_T as indicated in Figure 3.5, implying that the reservoir is somewhat run down during period $u + 1$. The amount of water saved for period $u + 2$ is indicated in the figure as the gap between consumption in period u and $u +$ 1. If the demand curve for period $u + 1$ is shifted to the right as indicated by the broken demand curve the consumption would be less than the inflow *CD* and the reservoir would be built up during period $u + 1$.

The optimal price cannot be lower than the price indicated by the intersection of the demand curves by the dotted horizontal line p_T^{min} for the figure to function. At this price level all available water will be used up in the two periods. The equilibrium price must be lower than the lowest choke price for period u in the figure, since we have assumed that there is positive consumption of electricity in all periods.

Scarcity in a period other than the terminal

We will now investigate what happens if the reservoir is emptied in other periods than the last one, i.e., we study a period $t + 1 < u$, and assume that the reservoir is emptied in that period, and furthermore assume that the price has been constant equal to p_T since the terminal period. Using conditions (3.5) and (3.6) we have for the two periods t and $t + 1$:

$$\begin{aligned}
&p_t(e_t^H) = \lambda_t \, (e_t^H > 0) \\
&-\lambda_t + \lambda_{t+1} - \gamma_t \le 0 \; (= 0 \text{ for } R_t > 0) \\
&p_{t+1}(e_{t+1}^H) = \lambda_{t+1} \, (e_{t+1}^H > 0) \\
&-\lambda_{t+1} + \lambda_{t+2} - \gamma_{t+1} \le 0 \; (= 0 \text{ for } R_{t+1} > 0)
\end{aligned} \tag{3.14}$$

The link with our optimal path story is that $\lambda_{t+2} = \lambda_T = p_T$. We will assume that there are no threats of overflow neither in period t nor period $t + 1$ implying $\gamma_t = \gamma_{t+1} = 0$. Furthermore, by assumption we have that $R_t > 0$ and $R_{t+1} = 0$. We assume strictly positive prices for all periods. Combining conditions and assumptions yields:

$$\begin{aligned}
&\lambda_t = \lambda_{t+1} > 0 \\
&p_t(e_t^H) = p_{t+1}(e_{t+1}^H) > 0 \\
&\lambda_{t+1} \ge \lambda_T > 0 \;\; (R_{t+1} = 0)
\end{aligned} \tag{3.15}$$

The normal situation would be to have strict inequality in the last condition: $\lambda_{t+1} > \lambda_T$. We can use Figure 3.5 as an illustration (setting $t = u$) assuming now that $0 < p_T < p_T^{\min}$. This price from the future is too low to influence the consumption of electricity in periods t and $t + 1$.The water allocation on the two periods is found by the intersection of the demand curves indicated by a vertical dotted line down from the intersection point to the bathtub floor. The price will be the same in the two periods as indicated by the dotted horizontal line through the intersection point of the two demand curves, actually the price $p_T^{\min}$. All the available water will be used up in period $t + 1$ since the water value in period $t + 1$ is higher than p_T. We note that the price in periods before this second scarcity period $t + 1$ will be higher than the price during the periods with neither overflow nor scarcity for the periods $t + 2, \ldots, T$, assuming neither overflow nor scarcity going backwards in time from $t + 1$.

Threat of overflow

The last case we will investigate is threat of overflow (reservoir completely filled) for a period $s < t$, where $t + 1$ is the first scarcity period after s going forward in time. Using condition (3.5) we have the general conditions for the two periods s and $s + 1$:

$$\begin{aligned} & p_s(e_s^H) = \lambda_s \; (e_s^H > 0) \\ & -\lambda_s + \lambda_{s+1} - \gamma_s \leq 0 \; (= 0 \text{ for } R_s > 0) \\ & p_{s+1}(e_{s+1}^H) = \lambda_{s+1} \; (e_{s+1}^H > 0) \\ & -\lambda_{s+1} + \lambda_{s+2} - \gamma_{s+1} \leq 0 \; (= 0 \text{ for } R_{s+1} > 0) \end{aligned} \tag{3.16}$$

The link with our optimal path story is that $\lambda_{s+2} = \lambda_t \; (= \lambda_{t+1}) > 0$ where $t > s + 2$. We assume that $R_s, R_{s+1} > 0, \gamma_s, \gamma_{s+1} > 0$. These conditions yield:

$$\begin{aligned} & p_s(e_s^H) = \lambda_s \; (e_s^H > 0) \\ & \lambda_s = \lambda_{s+1} - \gamma_s \; (R_s > 0) \\ & p_{s+1}(e_{s+1}^H) = \lambda_{s+1} \; (e_{s+1}^H > 0) \\ & \lambda_{s+1} \geq \lambda_t - \gamma_{s+1} \; (R_{s+1} \geq 0) \end{aligned} \tag{3.17}$$

The second equality in (3.17) follows from the Kuhn – Tucker condition in (3.5) when there is a positive amount of water in the reservoir. The shadow price on water λ_s is zero if there is actual overflow. This follows from the third condition (complementary slackness) in (3.5). If there is no spillage, as in our case with maximal manoeuvrability and the water is just maintained at the maximal level, the water value λ_s will typically be positive. In any case the water value λ_s is typically smaller than the water value λ_{s+1} for the next period because the shadow price on the upper reservoir constraint is typically positive.

To illustrate the possibility of overflow the total available water in a period must be greater than the reservoir storage capacity. In Figure 3.6 overflow threatens in period s if the price from period $t + 1$ is followed. The price for period s has to be lowered in order to avoid spilling, and the maximal reservoir filling *BC* is then saved to the next period $s + 1$, and *AB* is consumed in period s. In period $s + 1$ the price from the future, p_{t+1}, prevails, and somewhat more than the inflow *CD* is consumed, as indicated in the figure. This implies that the reservoir is run down in period $s + 1$ and somewhat less water than the full reservoir is left for period $s + 2$, as indicated in the figure.

In period s the shadow price on the reservoir constraint is the difference

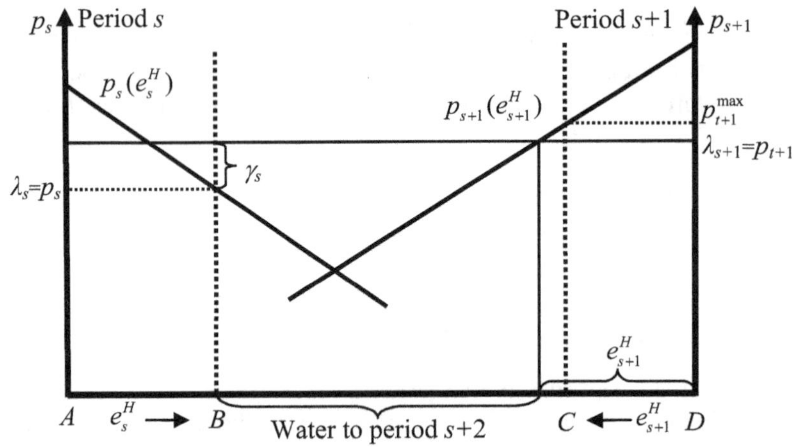

Figure 3.6. Threat of overflow.

between the price in period s and the price in period $s + 1$ that is equal to the price p_{t+1} given from the future. We may notice that for threat of overflow to occur in period s the price from the future cannot be lower than the price in period s necessary to generate enough demand, AB, to avoid spilling. A higher price from the future will still result in the same price for period s. This means that when we have an episode of threat of overflow the price from the future has no impact on the equilibrium price in the period with the threat of overflow. The link with future prices is broken. The management policy for periods in between the start and the period with threat of overflow does not have to take into consideration events beyond the period with the threat of overflow. However, the period with threat of overflow is endogenously determined in the planning problem, so the total problem has to be solved simultaneously.

There can be two consecutive periods with threat of overflow. If we consider the price from the future to be p_{t+1}^{max} indicated in the figure then the inflow in period $s + 1$ is just used up and the maximal reservoir filling is passed on to period $s + 2$. For higher future price than this level the price in period $s + 1$ cannot become higher without causing overflow. We would then also have a threat of overflow in period $s + 1$ and a difference between the period $s + 2$ price and the price p_{t+1} from the future equal to the shadow price on the reservoir capacity constraint in period $s + 1$. Each period of overflow will have its own price, thus a series of overflow periods can generate a sequence of price changes.

Output constraints

The load-duration curve shown in Figure 1.3 in Chapter 1 for Norway illustrates that power capacity may become a limiting factor even at an aggregated level. The spare capacity left at the historic peak in 2001 was only 8%. When the transmission system is not explicitly modelled and power and energy are not distinguished, then an upper constraint on the production during one period covers all these events at the aggregated level. We will call the constraint the production constraint in the following. It is stated as:

$$e_t^H \le \overline{e}^H \text{ , } t = 1,..,T \tag{3.18}$$

where $\overline{e}^H$ is without a time subscript since it is treated as a technical constraint. Sufficient power capacity means that

$$x_t^{\max} < \overline{e}^H \text{ , } t = 1,..,T \tag{3.19}$$

where $x_t^{\max}$ is the highest power demand, found close to the left axis of the load-duration curve in Figure 1.3 in Chapter 1. [We continue measuring the variables in (3.19) in kWh below.]

Inserting the production constraint (3.18) into the social planning problem (3.3) yields:

$$\begin{aligned}
&\max \sum_{t=1}^{T} \int_{z=0}^{x_t} p_t(z)\,dz \\
&\text{subject to} \\
&R_t \le R_{t-1} + w_t - e_t^H \\
&R_t \le \overline{R} \\
&e_t^H \le \overline{e}^H \\
&R_t, e_t^H \ge 0 \\
&T, w_t, R_o, \overline{R}, \overline{e}^H \text{ given}, R_T \text{ free}, t = 1,..,T
\end{aligned} \tag{3.20}$$

The Lagrangian for the problem is:

$$\begin{aligned}
L = &\sum_{t=1}^{T} \int_{z=0}^{e_t^H} p_t(z)\,dz \\
&-\sum_{t=1}^{T} \lambda_t (R_t - R_{t-1} - w_t + e_t^H)
\end{aligned}$$

$$-\sum_{t=1}^{T}\gamma_t(R_t-\overline{R})$$
$$-\sum_{t=1}^{T}\rho_t(e_t^H-\overline{e}^H) \tag{3.21}$$

Endogenous variables are e_t^H, R_t, λ_t, γ_t, ρ_t ($t = 1,\ldots,T$), and there are $5T$ variables in all. The necessary first-order conditions are:

$$\begin{aligned}
&\frac{\partial L}{\partial e_t^H}=p_t(e_t^H)-\lambda_t-\rho_t\le 0\ (=0 \text{ for } e_t^H>0)\\
&\frac{\partial L}{\partial R_t}=-\lambda_t+\lambda_{t+1}-\gamma_t\le 0\ (=0 \text{ for } R_t>0)\\
&\lambda_t\ge 0 (=0 \text{ for } R_t<R_{t-1}+w_t-e_t^H)\\
&\gamma_t\ge 0 (=0 \text{ for } R_t<\overline{R})\\
&\rho_t\ge 0 (=0 \text{ for } e_t^H<\overline{e}^H), \qquad t=1,..,T
\end{aligned} \tag{3.22}$$

There are $2T$ conditions in (3.22) and $3T$ more equations in (3.21), $5T$ in all. The manoeuvring of the system due to the production constraint now becomes an issue. Without the production constraint the system is perfectly manoeuvrable. But if there is too much inflow to the reservoir in a situation with a high level of reservoir filling, the production constraint may prevent enough water to be processed to avoid overflow. If the production constraint is effective then the water value is less than the social price according to the first condition in (3.22). The condition holds with equality, and constraining the processing of water implies that less is used than optimal without the constraint. The price will therefore have to rise. The water value measures the marginal cost of using water and is equal to the water value in the next period if the reservoir constraint is not binding and we do not have any scarcity situation in the period. The level of total demand will in general influence positively the occurrence of a binding production constraint. This may happen in peak load periods and be an additional reason for high prices.

There are two situations that can lead to the production constraint becoming binding: preventing overflow and trying to satisfy demand in a high social price period. The manoeuvrability of the system now depends on the minimum number of periods, t^o, it takes to empty the reservoir;

$$t^o = \min t \text{ such that } t\overline{e}^H \ge \overline{R}, \tag{3.23}$$

where t^o and t are integers. The higher this number is the less manoeuvrability. If the most favourable price regime lasts a number of periods less than t^o, either a full reservoir does not have to be accumulated before the high price periods, or it will be some water left in the reservoir after the high price regime.

Preventing overflow has to be planned for several periods before the actual threat of overflow if inflows are higher than the production capacity for some periods before the threat of overflow. The management task is to create enough space in the reservoir to contain the inflows without spilling water. Manoeuvrability, meaning the ability to run down the reservoir level, is present only for periods when production can exceed inflow: $e_t^H > w_t$. A certain combination of inflow patterns and production restriction may lead to a *locking-in* of water. This may happen if overflow is physically inevitable, as is the case if, starting with an empty reservoir; the inflows are such that the reservoir flows over in a later period although full production capacity has been used in all periods. Let the starting period be t' and the first overflow period be t''. The formal definition of a system lock-in situation is:

$$R_{t'} = 0, \sum_{t=t'}^{t''} w_t - (t'' - t' + 1)\bar{e}^H > \bar{R} \tag{3.24}$$

In Figure 3.7 the production constraint is dimensioned in such a way that the optimal amount of water without the constraint in period 1 can be processed, but not the water available for period 2 without meeting the

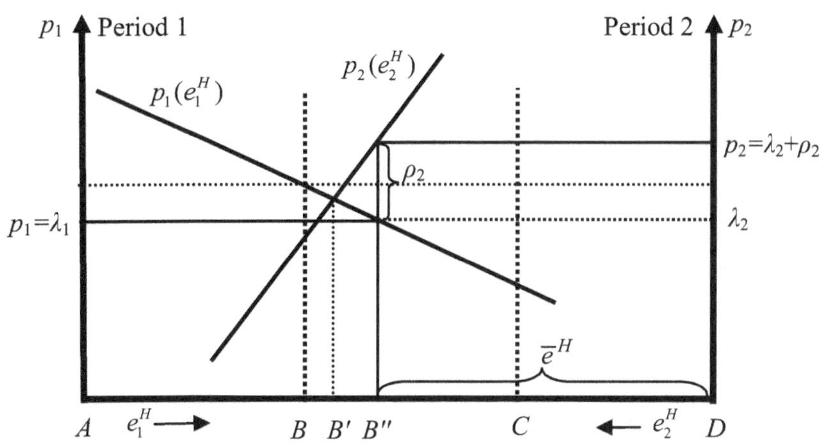

Figure 3.7. Production constraint binding in period 2.

constraint on the production capacity. The capacity constraint $\overline{e}^H$ is measured from the period 2 axis to the left and is shown by *DB''*. Without the production constraint the transfer of water from period 1 would be given by the intersection of the demand curves at *B'* on the horizontal axis. The prices would be equal as indicated by the thin horizontal dotted line through the intersection point of the demand curves. The amount *B'C* would have been transferred to period 2, but with the constraint only the smaller amount *B''*C is transferred. The use of water in period 1 is *AB''*, resulting in a lower price in period 1 than without the constraint, and a correspondingly higher price in period 2. We see from the water-value dynamics in (3.22) that the water values are the same for the two periods. More inflow in period 1 would result in more water processed in period 1, and the same would happen if there is a marginal increase in the inflow in period 2 since production in period 2 is constrained. The price hike in period 2 is caused by the production constraint.

By manipulating the storage lines and the line for the production constraint in the aggregate two-period model the production constraint may become active in the first period, as shown in Figure 3.8. The production constraint is now measured from *A* to the vertical line at *B'*. The maximal available water *AB* in period 2 after saving the full reservoir *BC* for period 2 is greater than the production constraint. By backwards induction and our general assumption of non-satiation of consumption, and specific assumption that the production constraint is not binding in period 2, we have that $p_2\,(e_2^{\,H}) = \lambda_2 > 0$. In period 1 we have assumed that the production con-

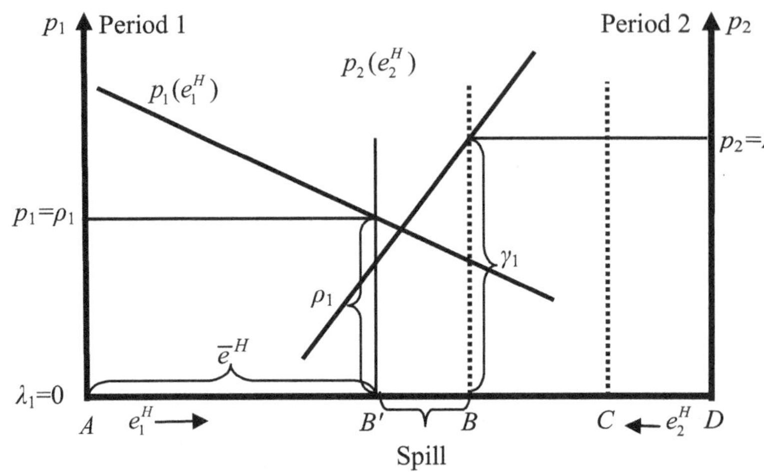

Figure 3.8. Production constraint binding in period 1. Locking-in of water.

straint becomes binding, i.e., $\rho_1 \geq 0$. The water-value dynamics does not involve this shadow price explicitly and yields $\lambda_1 = \lambda_2 - \gamma_1$ because the second condition in (3.22) holds with equality. The maximal reservoir *BC* is transferred to period 2. However, in the figure we have illustrated spilling in period 1, implying that the water value is zero, yielding $\lambda_1 = 0 \rightarrow \gamma_1 = \lambda_2$.The price in period 1 is then determined solely by the shadow price on the production constraint: $p_1\,(e_1{}^H) = \rho_1 > 0$.

The programming model assigns the extreme value of zero to the shadow price on stored water in period 1 while the production is evaluated to a positive social price. From the model point of view this is logical, because the accumulation of water during period 1 ends up with overflow and zero value is assigned to this flow. A marginal increase in accumulation of water in period 1 has zero value since the reservoir transferred to period 2 cannot become more than full. A zero water value is just a "go" signal for processing as much water as possible in period 1, but not a social evaluation of the actual consumption, which is positive and equal to the shadow price on the production constraint in period 1. But without the production restriction even a greater social value would have been created.

The two illustrations show us an important qualitative feature of the solution regarding prices and water values. Because of the production constraint there is now a potential difference between shadow value of stored water and value of processed water. A binding production constraint leads to difference between the value of water as stored water and as water being processed. The shadow-price dynamics in (3.22) only involve shadow prices related to the value of stored water, while the social price may now change between periods owing to the production constraint becoming binding and the condition of equality between supply and demand. One more cause of differences between the social prices has been identified.

Noncontrollable electricity generation

In most hydro systems power is also generated without having reservoirs that are relevant for the time unit of the analysis. This may be rivers, where what flows in must be produced continuously or else the water is lost. In Norway power from plants without storage possibilities constitutes about 30% of yearly production. The social planning problem including run-of-the-river power generation is:

$$\max \sum_{t=1}^{T} \int_{z=0}^{x_t} p_t(z)\,dz$$

subject to

$$
\begin{aligned}
&x_t = e_t^H + e_t^R \\
&R_t \le R_{t-1} + w_t - e_t^H \\
&R_t \le \overline{R} \\
&x_t, e_t^H, R_t \ge 0,\ t = 1,..,T \\
&T, e_t^R, R_o, \overline{R}, w_t \text{ given, } R_T \text{ free}
\end{aligned}
\tag{3.25}
$$

Here e_t^R is the electricity produced in period t by run-of-the-river (e_t^R = w_t^R) with zero production-dependent operating costs assumed. Energy is now supplied both based on using reservoirs and run-of-the-river so the *energy balance* is entered as a new constraint (the first one). The river water has to be processed as it comes in order to avoid losing the value. The reservoirs have to be used as buffers to absorb the river-flow fluctuations.

Since the energy balance has to hold with equality we can for simplicity substitute for x_t in the optimisation problem, yielding the following Lagrangian:

$$
\begin{aligned}
L = &\sum_{t=1}^{T} \int_{z=0}^{e_t^H + e_t^R} p_t(z)\,dz \\
&- \sum_{t=1}^{T} \lambda_t (R_t - R_{t-1} - w_t + e_t^H) \\
&- \sum_{t=1}^{T} \gamma_t (R_t - \overline{R})
\end{aligned}
\tag{3.26}
$$

The necessary first-order conditions are exactly of the same form as (3.5) for problem (3.3). Our standard assumption is that electricity is produced every period (but now it may be more realistic that demand for electricity may be satiated). If hydropower from reservoirs is used, then the price is equal to the water value. If we assume that hydro from reservoirs is produced every period, then demand for electricity is not satiated and we have the same situation as described by Equation (3.6) with e_t^H + e_t^R as the argument in the demand function.

This may be illustrated in a bathtub diagram by extending the "walls" with the run-of-the-river and shifting the demand schedules accordingly, as shown in Figure 3.9, which is an adaptation of Figure 3.3, in the case of a river flow in both periods. The river flow is added to the controllable hydro to the left and to the right of the old walls of the bathtub drawn as broken vertical lines. The demand curve for period 1 now has to be anchored

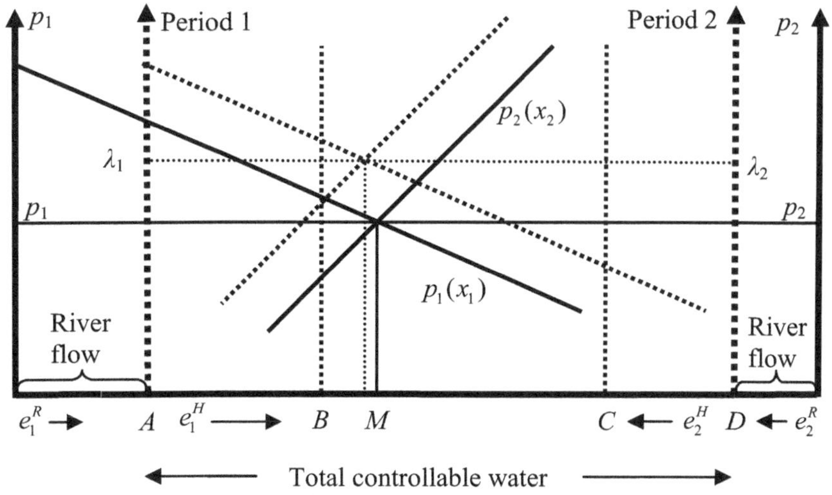

Figure 3.9. Run-of-the-river.
Controllable hydro only indicated by dotted lines.

on the river-extended wall marked with the solid vertical line to the left of the broken vertical line from *A*, and the demand curve for period 2 is anchored to the vertical line to the right of *D*. There are horizontal shifts of the demand curves (from the broken lines to the solid ones) equal to the river flow for both periods. The river flow in period 2 is smaller than the river flow in period 1. The part of the demands satisfied using controllable hydro are the *residual* demand curves. Neither before nor after adding the river flows is a maximal storage in period 1 needed in the example. The river flow in period 2 is smaller than in period 1, but the amount of stored water transferred to period 2 in order to keep water values equal is somewhat smaller, resulting in a greater increase in consumption in period 1 than in period 2. This can be seen by the slight shift to the right of point *M* (from storable hydro only to storable plus river flow) on the horizontal axis showing the distribution on the two periods. This is due to the fact that the demand curve for period 1 is more elastic than for period 2. This means that when the price decreases the consumption will increase relatively more in period 1.The river flows add to the total production so the common price for the two periods has to decrease. Other configurations are easy to accommodate in the bathtub diagram.

Uncontrollable river flows may cause extra variation in prices downwards in periods where these flows are substantial and demand low. Recalling the first-order condition for electricity produced by regulated hydropower we have:

$$p_t(e_t^H + e_t^R) - \lambda_t \leq 0 \ (= 0 \text{ if } e_t^H > 0), t = 1,..,T \tag{3.27}$$

For controllable hydropower not to be used, i.e., $e_t^H = 0$, we must have:

$$p_t(e_t^R) - \lambda_t \leq 0, t = 1,..,T \tag{3.28}$$

If the water value is higher than the social price we get using only unregulated flows, then storable water is saved for later periods. The current price is determined by inserting the actual river flow in the demand function $p_t = p_t\,(e_t^R)$. The price may even be driven down to zero.

Wind power may be treated in the same way as uncontrollable river flows. Electricity is generated by the wind blowing and moving the rotor blades that drive the generator (with a gear box in between). The cost of running a windmill is very much dominated by fixed maintenance and inspection costs not related to current production, so variable costs can be disregarded just as for hydropower. The energy balance equation with electricity generated by windmills is:

$$x_t = e_t^H + e_t^W, \tag{3.29}$$

where e_t^W is the windmill electricity. This relationship is then used in the problem formulation (3.25). The impact on optimal management of hydropower will be exactly as for run-of-the-river, so in Figure 3.9 "windmill electricity" may substitute for "river flow." Hydropower is used for satisfying the residual demand in an optimal way, obeying conditions of the type (3.27) with wind power substituted for river power. It is also straightforward to have both types of unregulated power at the same time.

In order to incorporate unregulated electricity production in a bathtub diagram we do not have to extend the bathtub walls as done in Figure 3.9. We can just focus on the demand curves for controllable hydropower as the residual demand curves and stay with a bathtub diagram like Figure 3.3, but shift the demand curves down with the size of the unregulated electricity generated in the periods as the horizontal distance.

Summing up causes of price variability

Running out of water and threat of overflow are the basic price-determining events. In our model formulation (3.3) this is captured by shadow prices on constraints becoming positive. A possible sequence of events is portrayed along the time axis in Figure 3.10 corresponding to the

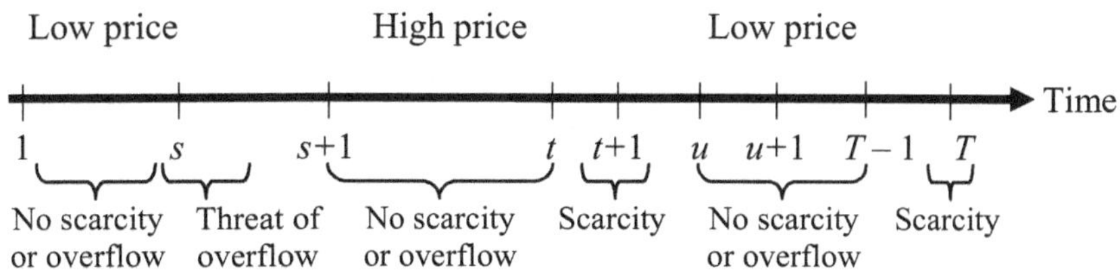

Figure 3.10. Main price-determining events.

cases we have investigated above. Starting backwards the terminal period is a scarcity period by assumption. Then there follows some periods with neither scarcity nor threat of overflow episodes covered by Figure 3.5. The price remains constant. Then a scarcity period is encountered and the price will jump up in this period, also illustrated in Figure 3.5, and continue to stay at this level when moving backwards in time provided we again have periods of neither scarcity or threat of overflow. If several scarcity periods occur without being interrupted by periods of threat of overflow, then the price is highest before the first incident of physical scarcity and the price is reduced successively each time a scarcity period is passed.

A period with threat of overflow will interrupt this sequence of price changes. The history after a period with threat of overflow does not count for the price formation in period leading up to the threat of overflow incident. Future influences on prices today are cancelled out by an incident of threat of overflow. Such an episode is assumed in Figure 3.10 in period s some periods after the second scarcity episode in period $t + 1$ (moving backwards). The price level of the scarcity period threatens overflow, and to avoid this, the price for period s has to be lowered, as illustrated in Figure 3.6. The shadow price on the upper capacity constraint of the reservoir is switched on. If the periods, going backwards on the time axis, again return to neither scarcity nor threat of overflow this lower price remains the price until the starting period.

If the optimal path of hydropower production and reservoir levels involves an interwoven pattern of scarcity periods and periods with threat of overflow, the price may cycle from higher values in periods after a threat of overflow episode to the next scarcity period and to a lower price after a scarcity period and until the next threat of overflow period. If we look to the left on the time axis in Figure 3.10 after a threat of overflow episode the connection to prices to the right on the time axis is completely broken. A succession of scarcity periods imply a building up of the price, being highest for the first scarcity period coming from the left on the time axis and then falling off after each scarcity period is passed until the last one. In this way our simple model may be able to generate a changing price pattern more in correspondence with what we observe.

The typical relation at the aggregate level between inflow and production in Norway was shown in Figure 1.4 in Chapter 1. From a management point of view, the acute problems arise at the end of the drawdown of the winter period and the filling up again during snow melting. In a few weeks the situation may change quickly from scarcity to threat of overflow for some hydropower plants. There may also be such episodes due to autumn rain as seen as smaller inflow peaks in Figure 1.4. However, at an aggregate level a typical yearly inflow cycle may generate only two major changes in the price regime. The three price regimes portrayed in Figure 3.10 may be for a yearly cycle and corresponding to start sometime during spring with a low reservoir level and then first facing threat of overflow at the peak of snow melting. There may be a second period of threat of overflow during autumn rains not indicated in the figure. The scarcity period may be in the next spring. Reservoirs will be drawn down during the winter and finally there may be no reason to hold back in early spring when temperature has risen and thawing has set in, but just to use up the water. The demand after the scarcity period must then be less than the immediate inflow (since the reservoir has been emptied) at the price charged, which is set reflecting the scarcity in the terminal period when the final emptying of the reservoir takes place. This price may be low, and even lower than the price we started with one year earlier. But this situation is a little artificial and created by our assumption of not looking beyond the planning horizon to the next snow melting. If the planning period is set to, e.g., two years the first spring encountered may still end with a scarcity period because room must be made available in the reservoirs for the coming snow melting. The price after the scarcity period will then prevail until the period with threat of overflow.

At the aggregated level the production constraint may become binding in high-demand periods. This will lead to an extra increase in the price in the high-demand periods. With reference to Figure 3.10 such an occurrence could be placed within the time interval before the first scarcity period.

Variability in electricity from river flows and windmills may explain price variations both when the reservoir constraint is not binding and when either the regulated hydro systems run up against constraints, or when the supply from the unregulated sources are so abundant that no hydropower is to be used, like night time during periods with snow melting or heavy rainfall, and persistent strong winds. Unregulated power may then especially explain short-term variation in prices hour by hour. Unregulated electricity will be used before stored water is produced, but running up against upper limits on reservoirs necessitates a higher current production and lower prices, thus contributing to price variation.

Determining quantities

In the previous section we only studied possible solutions for the prices. Addressing the determination of quantities it should, of course, be recognised that a solution is simultaneous in prices and quantities. We focus on quantities in this section only in order to obtain qualitative characterisations.

The development of the water in the reservoirs is keenly watched by the participants in the electricity market. The weekly developments of the aggregated reservoir level relative to the maximal level for Norway for 2005 together with the minimum and maximum relative levels for the period 1990-2005 are illustrated in Figure 3.11. The relative level changed from the lowest of 32% in week 16 (last week of April) to 92% in week 45 (second week of November). For the last weeks of the year the reservoir levels follow closely the maximum, and for all weeks the relative reservoir levels

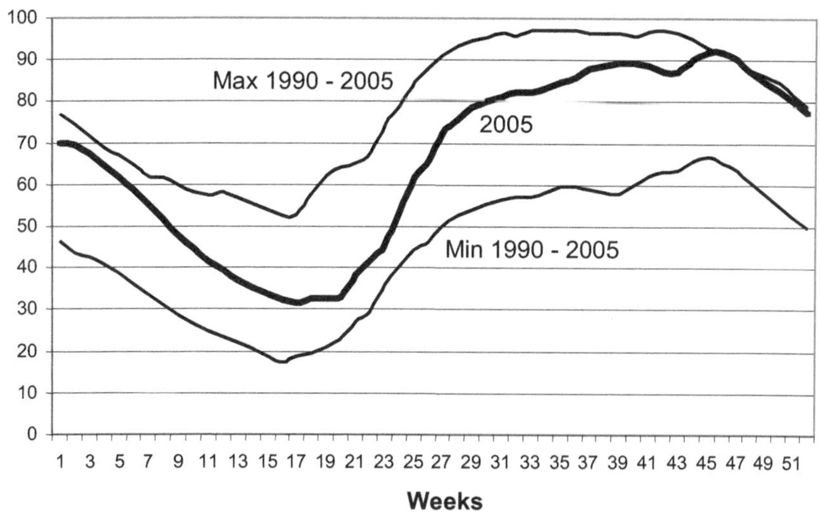

Figure 3.11. The weakly relative filling of the reservoirs. Norway year 2005. *Source: Nord Pool*

were comfortably above the minimum average for 1990-2005. The problematic period of scarcity is late April spring weeks with a minimum filling for the 15-year period of 17%. From August to November it is normal that the reservoirs fill up again to meet the winter demand, so in this period the problem is to manage without overflow. It will turn out below that the reservoir fillings have a crucial role to play on the quantity side.

Following again the principle of backwards induction we have that the solution for the production (production is always equal to consumption; for ease we will talk about production) in the terminal period is equal to the available water. Since the terminal value of the reservoir is free in the model (3.3) we assume that the reservoir will be emptied. The assumption of no satiation of demand is maintained. The following conditional solutions for production and prices are then obtained:

$$\begin{aligned} e_T^H &= \hat{R}_{T-1} + w_T \\ \lambda_T &= p_T(e_T^H) = p_T(\hat{R}_{T-1} + w_T) \end{aligned} \tag{3.30}$$

The solutions are conditional on the transfer of reservoir $\hat{R}_{T-1}$ from period $T-1$ to T.

Figure 3.12 shows that the range of water in the reservoir delivered from period T-1 is $(0, \max R_{T-1}) = (0, \bar{R})$, resulting in a range of $(w_T, w_T + \bar{R})$ for electricity production, and (OB, OA) for the shadow price λ_T on stored water. The optimal solutions (3.30) for electricity and shadow price on water depend on the amount of stored water transferred from period $T-1$.

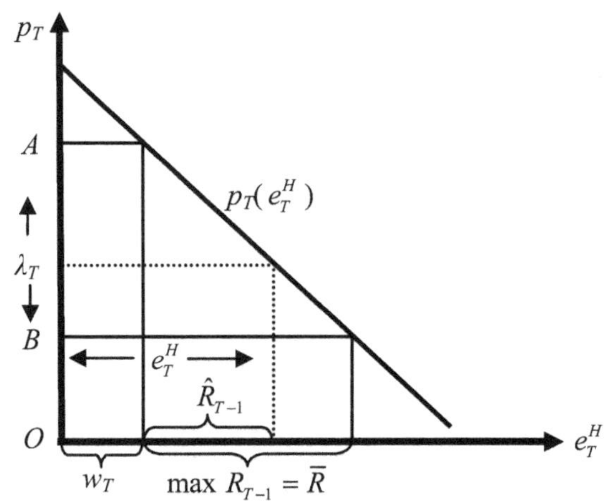

Figure 3.12. Backwards induction. Optimality in period T.

period $T-1$. The dotted lines indicate a possible (feasible) optimal solution.

In period $T-1$, e_T^H, λ_T are known given $\hat{R}_{T-1}$. The discussion of possible outcomes will be based on the events portrayed in Figure 3.10. The reservoir level is then assumed to take an interior value for period $T-1$. This assumption implies

$$\gamma_{T-1} = 0 \Rightarrow \lambda_{T-1} = \lambda_T, \tag{3.31}$$

using (3.5).

Figure 3.13 illustrates a feasible optimal solution for period $T-1$ contingent upon the possible solution for period T. The situation in the figure is such that more water is transferred from period $T-2$ to $T-1$ than from period $T-1$ to period T, i.e., consumption exceeds the inflow and the reservoir is run down in period $T-1$. A building up of the reservoir in period $T-1$ would imply that consumption is less than the inflow, and that more water is transferred to period T than was received from the end of period $T-2$.

The feasible solutions for the production levels will in general be in the interval

$$e_t^H \in \left[\max\,(0, R_{t-1} + w_t - \overline{R}), R_{t-1} + w_t\right],\ t = 1,..,T \tag{3.32}$$

Transferring the maximal amount to the next period yields the lowest production level in a period, and transferring zero yields the highest possi-

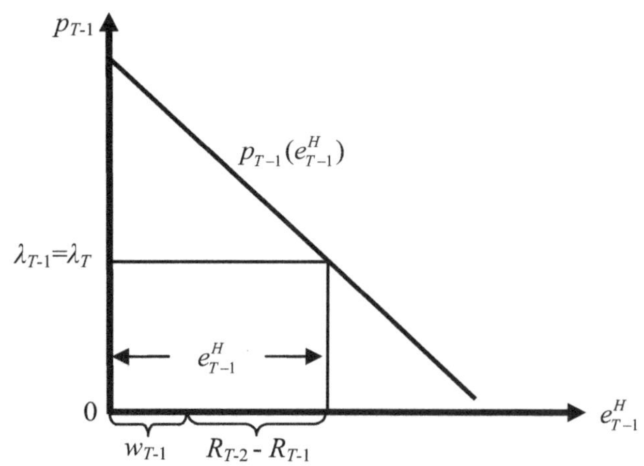

Figure 3.13. Feasible solution for period $T-1$ contingent on a solution for period T.

ble production level. Concerning the lower limit for electricity production in period t it should be noted that since electricity is non-negative, we have to exclude the possibility of a negative value if the available water is less than the maximal reservoir amount. It may happen in general in many periods that the available water is less than the reservoir limit, since the reservoir limit is without a period subscript and the same for all periods, and this limit will become relatively larger and larger compared with inflows as the period length is decreased. A reservoir limit of, e.g., 70% of the normal yearly inflow means that the inflow for an average week is less than 3% of the reservoir capacity, or put another way: for an average week the reservoir level at the end of the previous week must represent a filling of more than 97% for more than the reservoir content to be available. When the available water in a period exceeds the reservoir limit we cannot have a corner solution of transferring the total amount available to the next period, but must have an interior solution or the corner solution of transferring zero. When having the maximal transfer from a period to the next as a corner solution we will therefore have the situation that the available water in a period receiving a full reservoir necessarily exceeds the reservoir limit if the realised inflow is positive.

The shadow price on water for period $T-1$, determined by the water shadow price for period T, determines the electricity production via the demand function for period $T-1$:

$$\begin{aligned} &p_{T-1}(e^H_{T-1}) = \lambda_{T-1} = \lambda_T \Rightarrow \\ &e^H_{T-1} = p^{-1}_{T-1}(\lambda_{T-1}) = p^{-1}_{T-1}(\lambda_T) = p^{-1}_{T-1}(p_T(\hat{R}_{T-1} + w_T)), \end{aligned} \tag{3.33}$$

where p_{T-1}^{-1} is the inverse demand function. When the electricity production in period $T-1$ is determined we also have the solution for the transfer of water from period $T-2$ to period $T-1$ as a function of the transfer from period $T-1$ to T, using the water accumulation equation and inserting (3.33):

$$R_{T-2} = \hat{R}_{T-1} - w_{T-1} + e_{T-1} = \hat{R}_{T-1} - w_{T-1} + p^{-1}_{T-1}(p_T(\hat{R}_{T-1} + w_T)) \tag{3.34}$$

We can go backwards in this way right to period $t+1$ substituting successively from the equation of motion of the reservoir level. The solution for production in each period under the assumption $0 < \hat{R}_i < \overline{R}$ $(i = t+2,\ldots,T-1)$ as a function on the chosen level of reservoir filling at the end of period $T-1$ is:

$$e^H_i = p_i^{-1}(\lambda_i) = p_i^{-1}(\lambda_T) = p_i^{-1}(p_T(\hat{R}_{T-1} + w_T)),\ i = t+2,..,T-1,T \tag{3.35}$$

Concerning the reservoir level handed over to the next period the systematic substitution of the solution for the previous reservoir level as in (3.34) can be expressed in a general way by summing up available water in all the periods involved and the use of water:

$$\hat{R}_{t+1} + \sum_{i=t+2}^{T} w_i = \sum_{i=t+2}^{T} e_i^H \tag{3.36}$$

The level of the reservoir, $\hat{R}_{t+1}$, in period $t + 1$ is chosen from the feasible values. But assuming that there is a period of scarcity in $t + 1$ we know that nothing will be transferred to period $t + 2$, i.e., $\hat{R}_{t+1} = 0$. We also know that $R_T = 0$. Inserting this information into (3.36) and using the conditional solution (3.35) for production levels for the periods t +2,...,T – 1,T yield:

$$\begin{aligned} &\sum_{i=t+2}^{T} w_i = \sum_{i=t+2}^{T} e_i^H = \sum_{i=t+2}^{T} p_i^{-1}(p_T(\hat{R}_{T-1} + w_T)) \\ &\Rightarrow \hat{R}_{T-1} = \sum_{i=t+2}^{T-1} w_i - \sum_{i=t+2}^{T-1} p_i^{-1}(p_T(\hat{R}_{T-1} + w_T)) \end{aligned} \tag{3.37}$$

The solution (3.35) is used deriving the last expression above. This equation is only a function of the unknown level of transfer, $\hat{R}_{T-1}$, from period T – 1 to T and involves all inflows and all demand functions for the periods in question. Once we have this solution all the period production levels can be calculated from (3.35).

Moving backwards in time from $t + 1$ we have again periods with neither scarcity nor threat of overflow until period s, thus repeating the type of solutions above, but now with the transfer of water from period t to $t + 1$ as unknown:

$$\begin{aligned} &e_{t+1}^H = \hat{R}_t + w_{t+1} \\ &\lambda_{t+1} = p_{t+1}(e_{t+1}^H) = p_{t+1}(\hat{R}_t + w_{t+1}) \end{aligned} \tag{3.38}$$

Proceeding according to (3.35) updating (3.36) yields:

$$\hat{R}_s + \sum_{i=s+1}^{t+1} w_i = \sum_{i=s+1}^{t+1} e_i^H \tag{3.39}$$

Since we have assumed a threat of overflow in period s we know the transfer from period s to $s + 1$ is the maximal. Equation (3.37) can then be written:

$$\hat{R}_t = \bar{R} + \sum_{i=s+1}^{t} w_i - \sum_{i=s+1}^{t} p_i^{-1}(p_{t+1}(\hat{R}_t + w_{t+1})) \tag{3.40}$$

In this equation the only unknown is $\hat{R}_t$ so we can solve for this reservoir level. The solutions for the other reservoir levels for the periods $s + 1,..,t$ can then be found as above updating (3.37).

From period $s - 1$ backwards to the starting period we have again neither scarcity nor threat of overflow. The strategic unknown is now the reservoir transfer from period $s - 1$ to s. Repeating the reasoning above the solution for this level is found by solving for $\hat{R}_{s-1}$ from the following equation, remembering that R_o is known:

$$\hat{R}_{s-1} = R_o + \sum_{i=1}^{s-1} w_i - \sum_{i=1}^{s-1} p_i^{-1}(p_s(\hat{R}_{s-1} + w_s)) \tag{3.41}$$

The key role of the level of the reservoirs through time for backward induction may be one reason for the interest in the profession in diagrams for reservoir developments. Another reason may be more practical: it is change in reservoir levels or reaching certain levels that trigger actions as to amounts of release of water.

Chapter 4. Multiple Producers

Reservoir constraints

The reader may feel that assuming one hydro plant with one reservoir is limiting the realism of the model since there are over 700 hydropower plants in Norway, and a majority of them have reservoirs, 830 in all. We will therefore study the implications of several producers for the optimal allocation of water. We maintain the same assumptions as in Chapter 3 and regard only the upper constraint on the reservoirs in this section, but introduce more restrictions subsequently. Each plant is assigned one reservoir. A transmission system is not specified, and the plants operate independently, i.e., there are no "hydraulic couplings" as there will be between plants along the same river system. We will return to the issue in Chapter 7 and the latter issue in a section below. An important consequence of disregarding power, production or transmission constraints for any plant is that a plant can empty its reservoir during a single period. This can be defined as *perfect manoeuvrability* of the reservoirs. But we do not assume that inflows can be channelled to any reservoir. The inflows are reservoir or plant specific. The plants have in general different fabrication coefficients in their production functions (1.2) in Chapter 1, and the water-accumulation equation of the type (1.4) for each plant is deflated by the plant-specific fabrication coefficient, assuming no waste of water in production. We express formally all variables in kWh, although we will talk about water.

The planning problem is the same as (3.3) in Chapter 3, but now a subscript (j) for plant has to be introduced. A fixed number of N plants is assumed. We will also need a relationship connecting the amount consumed to the total amount produced. This is popularly termed the energy balance. The total amount consumed is x_t:

$$x_t = \sum_{j=1}^{N} e_{jt}^{H}, \; j = 1,..,N, \; t = 1,..,T \tag{4.1}$$

Electricity is a homogeneous good so it does not matter to the consumer who supplies the electricity. Plant supplies are just added together. The energy balance has to hold with equality due to the requirement of continuous physical equilibrium between production and consumption.

As in previous chapters the different consumer groups are still represented by a single aggregated demand function in total consumption for a period. The social planning problem is:

$$\begin{aligned}
&\max \sum_{t=1}^{T} \int_{z=0}^{x_t} p_t(z)\,dz \\
&\text{subject to} \\
&x_t = \sum_{j=1}^{N} e_{jt}^H \\
&R_{jt} \le R_{j,t-1} + w_{jt} - e_{jt}^H \\
&R_{jt} \le \overline{R}_j \\
&R_{jt}, x_t, e_{jt}^H \ge 0 \\
&T, w_{jt}, R_{jo}, \overline{R}_j \text{ given}, R_{jT} \text{ free}, \ j=1,..,N,\ t=1,..,T
\end{aligned} \tag{4.2}$$

The variables in the individual water accumulation equations are still measured in energy units (kWh), but plant-specific fabrication coefficients, a_j (j=1,..,N), are now used for the conversions. In order to simplify, substituting for total consumption from the energy balance into the objective function yields the Lagrangian:

$$\begin{aligned}
L = &\sum_{t=1}^{T} \int_{z=0}^{\sum_{j=1}^{N} e_{jt}^H} p_t(z)\,dz \\
&-\sum_{t=1}^{T}\sum_{j=1}^{N} \lambda_{jt}(R_{jt} - R_{j,t-1} - w_{jt} + e_{jt}^H) \\
&-\sum_{t=1}^{T}\sum_{j=1}^{N} \gamma_{jt}(R_{jt} - \overline{R}_j)
\end{aligned} \tag{4.3}$$

When operating with individual plants the shadow prices on the water accumulation constraints and the upper reservoir constraints are plant specific in the problem formulation. The necessary first-order conditions are:

$$\frac{\partial L}{\partial e_{jt}^H} = p_t\Big(\sum_{j=1}^{N} e_{jt}^H\Big) - \lambda_{jt} \le 0 \,(= 0 \text{ for } e_{jt}^H > 0)$$

$$\frac{\partial L}{\partial R_{jt}} = -\lambda_{jt} + \lambda_{j,t+1} - \gamma_{jt} \leq 0 (= 0 \text{ for } R_{jt} > 0)$$
$$\lambda_{jt} \geq 0 (= 0 \text{ for } R_{jt} < R_{j,t-1} + w_{jt} - e_{jt}^{H})$$
$$\gamma_{jt} \geq 0 (= 0 \text{ for } R_{jt} < \overline{R}_{j}) , \qquad t = 1,..,T , j = 1,..,N \tag{4.4}$$

Counting number of variables and independent equations in the system (4.3) - (4.4) there are $4TN$ endogenous variables (e_{jt}^{H}, R_{jt}, λ_{jt}, γ_{jt}), including $2NT$ individual plant level outputs and reservoir levels, and $2NT$ shadow prices, $TN + N$ exogenous variables ($w_{jt}, \overline{R}_{j}$), and the number of equations is $4TN$. However, as we shall see the structure of the conditions is such that we will not get unique solutions for all individual plant variables in general.

In order to simplify making qualitative interpretations possible, we assume that electricity is consumed in all periods to positive prices; $x_t > 0$, $p_t(x_t) > 0$ ($t = 1,\ldots,T$), implying that in each period at least one plant must have positive production of electricity. The first condition of (4.4) shows that a plant-specific water value may differ from the social price if the plant has zero production: $\lambda_{jt} \geq p_t(x_t)$ for $e_{jt}^{H} = 0$. The plant water value becomes zero if overflow occurs according to the complementary slackness condition. These are the two possibilities of plant water values deviating from the social price. However, overflow is obviously not optimal in our model.

Since electricity is a homogeneous good, the social price is independent of which plant that supplies the consumers. The existence of a common period price, and the optimality requirement that this price is equal to the individual plant water values if the plants are producing, is of crucial importance for understanding the optimal behaviour of the system. If a plant is to be used, the water value in the periods in which it is used has to be equal to the social price for the periods in question. Furthermore, other plants having positive production in the same periods must then also face the common prices.

As to the shadow price on the reservoir constraint, it measures in general the increase in the objective function of a marginal increase in the reservoir of plant j for a period. The shadow price on the upper reservoir constraint becomes zero if the constraint is not binding. If there is a threat of overflow in a period the dynamic shadow-price equation in (4.4) holds with equality. We see that a positive value of the shadow price can be realised only if the plant produces in the period after the threat of overflow, since the shadow price must be the difference between the plant's water value in the period after the one we are considering and the period in question. But assuming the inflow is positive in the period after the threat of overflow production has to be positive to avoid spilling. Then the water value becomes equal to the price. But this is the same situation for all plants since the social prices are common. If there is a price difference between the periods there cannot be a plant-specific shadow value on the reservoir constraint in the first period. Therefore the shadow-price dynamics, as stated in the second equation in (4.4), will be valid only for common values for all plants for the water values and the shadow prices on the upper reservoir constraints. The implication for the solution of the problem (4.2) is that the *TN* conditions in the second equation of (4.4) reduce to *T* conditions. We can only obtain unique solutions for the *aggregate* production in any period, but not solutions for the allocation of this production on individual plants.

We can see this by using the backwards-induction principle. Assuming that demand is not satiated and that all reservoirs are emptied in the terminal period *T*, due to the free terminal condition, we get:

$$-\lambda_{jT} - 0 \leq 0 \Rightarrow \lambda_{jT} = p_T(x_T) > 0, j = 1,..,N \tag{4.5}$$

The equality follows from the assumption that all units are producing electricity in the last period (at least the inflows w_{jT}) and that the market price is positive due to non-satiation. But the condition above is not specific to plant *j*, but applies to all plants. In the optimal solution all plants are assigned the *same* water value in the last period and the total production of electricity is $\sum_{j=1}^{N}(R_{j,T-1} + w_{jT})$. All water in the reservoirs carried over to period *T* is used up together with the inflows in the last period.

For period $T - 1$ the process is repeated. Without overflow at any plant or any plant emptying its reservoir all plants are again facing the same water values according to the second equation in (4.4) and the price must be the same as for period *T* and common to all plants. We can go backwards to period 1 and get the same result. However, we have to check how the system price can change and what happens when there are corner solutions for individual reservoirs and plants.

The nature of indeterminacy is illustrated in Figure 4.1. The social price is p_t and the demand function $p_t(x_t)$. At the given price the demand determines the total production from the plants, *OA*, as indicated in the figure. However, it does not matter for the optimal solution which plants that contributes as long as the plants have the optimal amounts in the reservoirs when the system price changes. Such a price change is illustrated for period $t + 1$, representing the higher demand with the broken downward-

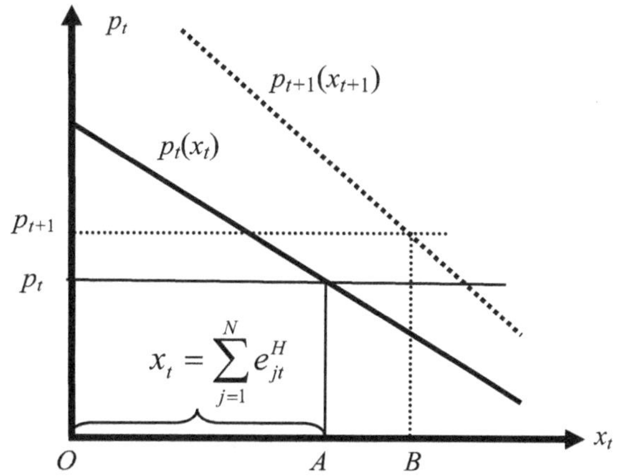

Figure 4.1. The nature of the optimal solution.

sloping line. The shift in the demand requires a production level to keep the period t price that it is not optimal to sustain, or not even feasible. The best that can be done is to have the production level OB, resulting in a higher period price indicated by the horizontal dotted price line. According to the explanation of price changes in the aggregated system in Chapter 3 the price increases when it is impossible to transfer enough water to sustain the same price, or to put it another way; the price in period t is lower due to the limit on the water that can be transferred to period $t + 1$, and more water than socially desirable without storage constraints has to be used in period t. For the reservoirs that it is physically possible to bring up to full level we must have in period t that $R_{jt} = \overline{R}_j$ for all j. The management problem is that $\sum_{j=1}^{N} \overline{R}_j$ is too small to keep the same price in high-demand periods as in low-demand ones.

Hveding's conjecture

In Chapter 3 episodes leading to price changes were investigated. Threat of overflow and emptying the reservoir were the price-determining events. In a multiplant model interesting questions are if, and how, this pattern is repeated. Specifically, may one plant have an overflow while none of the others have, and may one plant empty its reservoir and none of the others?

We will investigate the situation of overflow first and use the notation in Figure 3.10 in Chapter 3, where s was used for the period with a possibility of overflow. But actual overflow means that the water value is zero according to the complementary slackness condition in the third line of (4.4). Since the price is positive, by definition it cannot be optimal to have overflow for a plant alone. In fact, we cannot have overflow for any plant in the optimal plan since this is pure waste and we operate with perfect manoeuvrability of reservoirs and non-satiation of demand.

The next step is to investigate the case of threat of overflow, but no actual spilling of water. This situation means that $R_{js} = R_{j,s-1} + w_{js} - e^H_{js} = \overline{R}_j$. It would be rather arbitrary that it is optimal to keep this balance without drawing some water in period s, i.e., $e_{js}^H > 0$. Producing implies $\lambda_{js} = p_s(x_s) > 0$. Concerning period $s+1$ it is assumed that we have a positive price for this period, too; therefore we have from the shadow-price dynamics of (4.4): $-\lambda_{js} + \lambda_{j,s+1} - \gamma_{js} = 0$. Different from the situation in Chapter 3, we will here first assume that the social prices are the same for periods s and $s + 1$. Since there is a positive amount of water in the reservoir at the end of period s the dynamic equation for the shadow prices for plant j holds with equality. We will furthermore assume that the reservoir is below its limit in period $s + 1$. But the water values for period s and $s + 1$ must then be equal since the prices are equal by assumption. This means that if plant j is to face threat of overflow in period s, but not in period $s + 1$, then the shadow price on the reservoir constraint, γ_{js}, has to be zero. The conclusion is that, if we look at the *physical* situation of a reservoir, it is possible to have a threat of overflow at one reservoir only. But then the shadow price on the reservoir constraint must be zero. This is a possibility according to the Kuhn – Tucker condition, but this situation is seldom assumed or encountered by economists. This implies that the social objective function is not influenced by a situation of threat of overflow at one plant only in the interior of time intervals with the same social price. But there is one period when the shadow price becomes positive, and that is the period immediately preceding a price increase. After all the objective function must be positively influenced by an increase in the reservoir capacity $\overline{R}_j$. It may seem a rather arbitrary situation to have a threat of overflow and a zero shadow price at the same time. Seen from the shadow-price side the picture is simpler: isolated periods of threat of overflow for a single plant with a zero shadow price cannot be identified, but they do not matter for the value of the social objective function. During an interval of equal social prices the contribution of water to satisfy total demand may come from plants running a full reservoir (without spilling). However, these plants are not rewarded particularly for doing this (except in the period preceding a price rise). The water values remain equal to the social price. (Notice that if a plant within an interval with the same social price is run with a full reservoir for several consecutive periods, then the current inflow cannot be stored and becomes similar to a run-of-the-river flow.)

Let us now assume that $p_s(x_s) < p_{s+1}(x_{s+1})$. Since producing plants' water values are all equal to the social price for the same period such a price difference is possible only if *all* plants producing in both periods face a threat of overflow in period s. If all the plants face a threat of overflow in period s, but none in period $s+1$, we have the situation described in Chapter 3 for the aggregated system. All the plants face the same price for each period, implying the water values are equalised across plants, $p_s(x_s) = \lambda_{js}$, $p_{s+1}(x_{s+1}) = \lambda_{j,s+1}$, $j = 1,\dots,N$. According to (4.4) we then have $\gamma_{js} = p_{s+1}(x_{s+1}) - p_s(x_s)$. The shadow prices of the plant reservoir constraints are all equal for plants reaching the constraints.

But in the optimal plan we may also have plants that have not reached the reservoir constraint in period s even if they have accumulated water from the start. We have to investigate this possibility. Let us start with checking if one plant may accumulate water while all the other plants have filled up their reservoirs. Let us now first assume that the prices are the same again for period s and $s+1$. We know that zero production in period s implies that $\lambda_{js} \geq p_s(x_s)$, and that the shadow price on the upper reservoir constraint is zero since the reservoir is still not full by assumption. But then we get from the shadow-price dynamic equation that $(-\lambda_{js} + \lambda_{j,s+1}) = 0$. Such an accumulation episode is possible only if the water values are equal for the two periods. If plant j is producing in period $s+1$, the water value of the plant will equal the price. This implies that the water value in period s when the plant is not producing cannot be higher than the price assumed to be the same for period s and $s+1$. Again this is mathematically possible, but unusual according to standard interpretations by economists. Pure accumulation may take place in some plants and not others due to the balancing of total demand and total supply period for period.

Accumulation may continue in period $s+1$ and for more periods until water is processed. But then we have that the social price in the period production starts again determines all the shadow prices back in time, and furthermore, the price in the period before accumulation started up must be the same as the price in the period production resumes, assuming no threat of overflow in the former period. Let us now assume again that the price in period $s+1$ is higher than in period s. A plausible situation may be that a plant with a huge enough reservoir (or the inflow is small compared with the size of the reservoir) may not physically be able to reach the reservoir constraint in period s, i.e., $\sum_{t=1}^{s} w_{jt} + R_{jo} < \overline{R}_j$. But then it is probably the situation that it is optimal to accumulate right from the first period. This may be the case for a few reservoirs designed to take years to fill up and serving as insurance against especially dry years; *multiyear reservoirs*. The first period such a reservoir will be used will then determine the water value in all previous periods right back to the start. Remember that the model is deterministic. Whether such a plant will used in period $s + 1$ then depends on whether there will be a future period with an even higher price than in period $s + 1$ within the horizon T such that the plant can continue accumulating without meeting the reservoir constraint. In this case there will be no production in period $s + 1$. However, if there is no such period within the planning horizon a multiperiod reservoir will be drawn down sooner or later even though it has never been filled up completely. As pointed out above, the reservoir may come on and off more than one time, but this demands that the prices for the periods the plant is producing must be the same.

The conclusion is that in the multiplant model an increase in price from period s to period $s + 1$ typically requires that plants that physically cannot reach the reservoir limit in period s, have no production in period s, i.e., they are accumulating water. The equilibrium between supply and demand determines how many plants are involved in pure accumulation.

The other extreme situation is that plant j empties its reservoir in period $t + 1$, but not the other plants. Let us assume the relevant situation is that the prices are equal for two periods, t and $t + 1$. The first condition in (4.4) yields $\lambda_{j,s+1} = p_{t+1}(x_{t+1})$ since plant j has positive production. The second condition in (4.4) now yields $(- \lambda_{j,s+1} + \lambda_{j,s+2}) \leq 0$, $R_{j,t+1} = 0$ since the shadow price on the reservoir constraint in period $t + 1$ is zero. Assuming strict inequality we have for plant j that it is required that $p_{t+2}(x_{t+2}) > p_{t+1}(x_{t+1})$, while the condition for the other plants yields $p_{t+2}(x_{t+2}) = p_{t+1}(x_{t+1})$. But this is a contradiction. We conclude that in the regular case all reservoirs have to be emptied at the same time for the plan to be optimal. But note that the inequality involved is not strict, so it may be optimal for plants to empty their reservoirs before others. This latter case requires that the water value for plant j remains the same for the two periods, implying that the value of the social objective function may remain the same. We have a similar dichotomy as for the case of overflow above: the shadow prices tell a simple story of no economic impact of scarcity as long as the water values remain equal across plants and across time, while concerning the physical situation a plant may empty its reservoir before others, but then this should not influence the value of the social objective function for an optimal plan.

If all the plants face an episode of going empty in period $t + 1$, but in the immediate preceding or following periods they are in between scarcity and upper reservoir limits, we have the situation described in Chapter 3 for the aggregated system. All the plants face the same price for each period since they are producing, implying the water values are equalised across plants. The shadow price on the reservoir constraint in period $t+1$ is then zero. We then have $(-\lambda_{j,t+1} + \lambda_{j,t+2}) \leq 0$. Adopting strict inequality as the regular case we must have $p_{t+1}(x_{t+1}) > p_{t+2}(x_{t+2})$ according to the two first conditions in (4.4). It would therefore be arbitrary for all the water values to become equal for the two time periods.

To check if one plant may not empty its reservoir in a period $t + 1$ while all the other plants do, let us assume that the social price for period $t + 1$ is higher than for $t + 2$ in accordance with the example in Figure 3.10 in Chapter 3. The water values for this plant must then be equal for the period $t + 1$ and $t + 2$ for the social planner not to empty the reservoir for this plant also. But this leads to a contradiction. Thus this constellation cannot be a part of an optimal plan.

The reasoning above leads to the following result for the multiplant model (4.2) under the maintained assumptions:

> **Hveding's conjecture:** *In the case of many independent hydropower plants with one limited reservoir each, assuming perfect manoeuvrability of reservoirs, but plant-specific inflows, the plants can be regarded as a single aggregate plant and the reservoirs can be regarded as a single aggregate reservoir when finding the social optimal solution for operating the hydropower system.*

In Hveding's words:

> ...no single reservoir is overflowing before all reservoirs are filled up, and ... no single reservoir is empty before all are empty (Hveding, 1968, p. 131).

It is straightforward to aggregate all reservoirs as long as water has the same shadow price, and this also holds for aggregating reservoir constraints when they apply in the same period and have the same shadow price. When reservoir constraints are binding for other periods we noted that their shadow prices were zero. The individual reservoirs may then all be utilised in the same fashion, *as if* there is only one reservoir, with the qualifications elaborated upon above. This is a result of important practical value since it may simplify greatly the modelling effort. The results about price movement studied in Chapter 3 for one plant and one reservoir are all valid also for the multiplant case. The assumption of plant-specific inflows is crucial for the possible difference between movement of aggregate prices and individual water values. Without this assumption Hveding's conjecture would be rather straightforward, but would not serve as fruitfully as a benchmark for the management of individual plants.

As mentioned above, the model (4.3)-(4.4) does not determine the individual water release profiles of the plants. What we can say about individual profiles is that plants should, if possible, be brought up to full reservoirs in the same period and brought to empty conditions in the same period (with the exceptions mentioned above). Aggregation to meet market demand in between price-changing periods may involve varying contributions from the plants. The plant reservoirs may have different characteristics as to patterns of seasonal inflow and storage capacities both absolute and relative, although they are perfectly manoeuvrable. The possibility of such differences is allowed under our assumptions.

One way of thinking about adjustment of individual plants is to disregard the process of finding the optimal period social prices and just take them as given. Then each plant should be operated such that as much value as possible is created from the water. The price may fluctuate over the season more or less, as investigated for the total system in Chapter 3. A plant should accumulate water to meet the high price periods that will be common to all plants. A plant with good storage possibilities should then be left to accumulate water compared with a plant with little storage possibility. In the running up to the high price period plants with lower storage possibilities will therefore contribute more to the current production. Plants with good storage possibilities may at the extreme produce only in the peak period with the highest price (remember that per assumption plants can process all stored water in a single period). For such plants the first condition in (4.4) may hold with an inequality, i.e., the water value is higher than the current period price, until the high price period. But the total capacity of such plants may be higher than the market demand at the price in question, so the pattern of use of individual plants may still differ. The plants that fill up again more rapidly may be required to run down their reservoirs correspondingly more frequently. We know from Chapter 3 that if overflow threatens, as it may during periods leading up to reservoirs becoming full, then the price level may remain lower than the eventual peak price level for many periods. In order to be ready for the peak price period, plants may be run at levels of maximal storage capacity during these lower price periods. Then current inflows have to be processed as run-of-the-river plants.

Hveding's conjecture does not say necessarily that all plants must face the upper reservoir constraint always at the same time. But all plants that can physically do so have to hit the reservoir constraint in the period before a price increase. Some plants with low rates of inflows and large reservoirs may still accumulate water while others keep full reservoirs. But the shadow price on the reservoir capacity of the former type of plant will be zero. This implies that when aggregating such a reservoir capacity the real capacity limit should not be added, but only the actual maximal use, $\max_t R_{jt}$, $t = 1,\ldots,T$. The point is that the slow accumulators should be in the best possible position when the high price period arrives. But notice the restrictions on aggregate prices mentioned above for such episodes to be part of an optimal plan.

Hveding's conjecture justifies using a single plant-single reservoir model, but the conjecture does not give us a detailed plan for how to operate individual plants in a complex system. Specifically, the plants should not be required to fill up the reservoirs and draw them down on a strict equal-percentage basis, although this may serve as a simplifying benchmark if the relationship between inflow and reservoir capacity is not too different.

Optimal management of the system implies that price differences are kept at a minimum. Our social planner sees to this, although we can only indicate qualitatively what optimal utilisation of individual plants may entail. The interesting and intriguing story is whether a decentralised market can find the optimal patterns of individual plant use. It is important to understand that a well-functioning market in a technical sense is not automatically mimicking a social optimal solution of the type following from solving the model (4.2). We return to this issue in Chapter 10.

Output constraints

Hveding's conjecture may not hold strictly if more of the constraints entered in Table 3.1 in Chapter 3 are introduced. The constraints may be so demanding to fulfil, especially with a fine time resolution, that some reservoirs may experience overflow and some may be emptied before others. This has to be investigated more closely, starting with production constraints.

In order to satisfy the energy demand at the rate shown by the left-hand part of the load-duration curve in Figure 1.3 in Chapter 1, the system must have sufficient power capacity. When we do not model explicitly the transmission system and do not distinguish between power and energy, then an upper constraint on the production during one period for each plant covers all these events. (For a finer time resolution when these latter constraints can be identified only one of the constraints will in general be binding at the same time). The constraint for each plant is:

$$e_{jt}^{H} \leq \overline{e}_{j}^{H} \quad , j = 1,..,N,\ t = 1,..,T \tag{4.6}$$

where $\overline{e}_j^H$ is the upper power, production or transmission constraint for plant j. In Chapter 3 such a constraint was used for the whole system. However, each plant faces this constraint making the model more realistic when including it in the model. Sufficient system power capacity now means that

$$x_t^{\max} < \sum_{j=1}^{N} \overline{e}_j^H , \; t = 1,..,T , \tag{4.7}$$

where $x_t^{\max}$ is the highest power demand, found close to the left axis of the load-duration curve in Figure 1.3 in Chapter 1. However, locking-in of water at individual reservoirs, as mentioned in Chapter 3 for the whole system, may imply that individual plant capacities cannot simply be added as in (4.7) when calculating the system production capacity. The system capacity may be smaller. We will return to this topic below.

The social planning problem is:

$$\max \sum_{t=1}^{T} \int_{z=0}^{x_t} p_t(z)\,dz$$

subject to

$$\begin{aligned}
& x_t = \sum_{j=1}^{N} e_{jt}^H , \; t = 1,..,T \\
& R_{jt} \le R_{j,t-1} + w_{jt} - e_{jt}^H \\
& R_{jt} \le \overline{R}_j \\
& e_{jt}^H \le \overline{e}_j^H \\
& R_{jt}, x_t, e_{jt}^H \ge 0 \\
& T, w_{jt}, R_{jo}, \overline{R}_j, \overline{e}_j^H \text{ given, } R_{jT} \text{ free, } j = 1,..,N, \; t = 1,..,T
\end{aligned} \tag{4.8}$$

The fourth constraint above is the new one on the upper level of production. It is reasonable to assume that this limit is independent of the period since it is a technical constraint. Constraining the rate of production means that it may take more than one period to empty the reservoir when it is full. This plant-specific number of periods, t_j^o, is simply given by the minimum integer number equal or greater than $\overline{R}_j / \overline{e}_j^H$ and is a straightforward generalisation of (3.23):

$$t_j^o = \min t_j \text{ such that } t_j \overline{e}_j^H \ge \overline{R}_j , \; j = 1,..,N , \tag{4.9}$$

where t_j, t_j^o are integers. To run a model without an upper restriction on production as in the previous section is the same as assuming that $t_j^o = 1$. This *plant-specific minimum emptying time* give information about the manoeuvrability of the plant: maximal manoeuvrability is obtained when $t_j^o = 1$, and then manoeuvrability decreases as minimum emptying time increases. A *plant-specific manoeuvrability index*, m_j, may be defined as the inverse of the minimum emptying time giving the most flexible situation, index value 1, and increasing inflexibility toward index value zero:

$$m_j = \frac{1}{t_j^o},\ m_j \in (0,1],\ j = 1,..,N \tag{4.10}$$

The value of the manoeuvrability index will tell the planner when care has to be exercised as to how much water should be accumulated before high-price periods. A low value of m_j may imply that there is plenty of water left when the high-price periods are over. This may be a problem for two reasons: periods with seasonally higher inflows may be approaching and a low level of the reservoir is necessary in order to contain the inflows in the reservoir, and prices may be lower after the high-price periods than before. In the latter case more water should then have been used before the high-price periods. Plants with high values of the index should accumulate maximally in front of high-price periods. Since the model is deterministic, the necessary information for optimal management is available to the planner.

Substituting for total consumption from the energy balance in the objective function in (4.8), the Lagrangian is:

$$\begin{aligned} L = & \sum_{t=1}^{T} \int_{z=0}^{\sum_{j=1}^{N} e_{jt}^H} p_t(z)\,dz \\ & - \sum_{t=1}^{T} \sum_{j=1}^{N} \lambda_{jt} (R_{jt} - R_{j,t-1} - w_{jt} + e_{jt}^H) \\ & - \sum_{t=1}^{T} \sum_{j=1}^{N} \gamma_{jt} (R_{jt} - \overline{R}_j) \\ & - \sum_{t=1}^{T} \sum_{j=1}^{N} \rho_{jt} (e_{jt}^H - \overline{e}_j^H) \end{aligned} \tag{4.11}$$

The necessary first-order conditions are:

$$\begin{aligned}
&\frac{\partial L}{\partial e_{jt}^H} = p_t(\sum_{j=1}^{N} e_{jt}^H) - \lambda_{jt} - \rho_{jt} \le 0 (= 0 \text{ for } e_{jt}^H > 0)\\
&\frac{\partial L}{\partial R_{jt}} = -\lambda_{jt} + \lambda_{j,t+1} - \gamma_{jt} \le 0 (= 0 \text{ for } R_{jt} > 0)\\
&\lambda_{jt} \ge 0 (= 0 \text{ for } R_{jt} < R_{j,t-1} + w_{jt} - e_{jt}^H)\\
&\gamma_{jt} \ge 0 (= 0 \text{ for } R_{jt} < \overline{R}_j)\\
&\rho_{jt} \ge 0 (= 0 \text{ for } e_{jt}^H < \overline{e}_j^H)
\end{aligned} \tag{4.12}$$

The individual reservoirs differ in capacity and inflow characteristics, and the stations differ in production (power) capacity relative to size of reservoir and inflow characteristics. Therefore the manoeuvring of the stations and the reservoirs may differ. The manoeuvring would be to avoid spilling water, since this will typically serve the objective of maximising consumer plus producer surplus.

We will only discuss solutions when the production constraint is binding, since non-binding constraint was discussed in the previous section.

The social price is common to all units, but when production constraints are binding the individual water values may no longer be the same across plants in the optimal solution. The water value becomes plant specific and is less than the period's social price, according to the first condition in (4.12). The condition must hold with equality since production is positive. As in Chapter 3, when studying the aggregated production constraint, there is a separation between determination of water values and determination of social period prices that all plants face in common.

We will look at three possibilities concerning two consecutive prices:

i) $p_t > p_{t+1}$
ii) $p_t = p_{t+1}$
iii) $p_t < p_{t+1}$

Constraining the amount of water that can be processed implies that more is kept in the reservoir than optimal without the constraint. The optimal value of the objective function is reduced and consequently the water value will be smaller than in the unconstrained case. The water value is less than the current price since more water cannot be processed in the current period even if the reservoir amount is marginally increased (through increased transfer from the previous period or increased inflow). According to the dynamic shadow-price equation in (4.12), assuming no threat of overflow in the current period, the water value in the current period is equal to the water value in the next period. Assuming a non-binding production constraint for our plant in the next period implies that for this constellation to be part of an optimal plan, the social price in the next period must be *smaller* than the current price. The logic is that if more water in the reservoir is available in the current period it cannot be used in that period, but in the next one. For the water value in the current period to be lower than the period price the price in the next period has to be lower.

It may be the case that the production constraint is binding for several periods. Assuming that the reservoir constraint is not binding, we have that the water value will be the same for these periods, and equal to the social price in the first period with a non-binding production constraint. This price must then be lower than all the prices for the preceding periods with binding production constraints for the shadow prices on these constraints to become positive.

Assume that the prices are the same for the two periods, and furthermore that the reservoir constraint is not binding in any period, and that the production constraint for our plant is not binding in the next period. Then the shadow price on the production constraint has to be zero in period t even though the constraint is binding. The logic is that since the prices are the same, there is no increase in the objective function for the aggregated system of relaxing the constraint in the first period for a single plant. More electricity can be obtained from plants not being constrained.

Entertaining the same assumptions as above, the third situation with the price in period t being higher than the price in period $t + 1$, is not consistent with those assumptions. The effect of a binding production constraint in period t cannot be to increase the price in period $t + 1$ when there is no binding constraint, and by assumption the reservoir constraint is not binding in period t, ruling out an increase in the water value.

The level of total demand will in general influence positively the number of upper constraints on reservoirs that would become binding since more water in the aggregate is needed. If the social price were to be changed due to binding restrictions it has to be to a higher level in period t. The more binding constraints the higher the period social price may become. Notice that binding production constraints in a period is now an event that may change prices between periods even though water values remain the same. This is illustrated in Figure 3.7 in Chapter 3 for the aggregated system.

It is now not the case that we must have all production constraints binding for the same period at least once, as was the case when price changes are only driven by changes in water values in the previous section. The number of binding production constraints may be said to be demand-driven. It is only if demand should be so high, perhaps due to unusually cold weather on a winter day, that the total system capacity may become so strained that all production constraints are reached.

As in the aggregated case in Chapter 3 there are two situations that can lead to production constraints becoming binding: preventing overflow and trying to satisfy demand in a high social-price period. The manoeuvrability of a plant now depends on the number of periods, t_j^o, it takes to empty the reservoir; the higher this number the less manoeuvrability according to the plant-specific manoeuvrability index, m_j. If the high-price regime lasts a number of periods less than t_j^o, either the plant does not have to accumulate a full reservoir before the price periods, or it will have some water left in the reservoir after the high price regime. The impact of a production constraint on a multiyear reservoir may be to stop pure accumulation sooner and start producing if the production constraint prevents all available water to be processed in the high-price period.

Preventing overflow has to be planned for several periods before the actual threat of overflow if inflows are higher than the production capacity for some periods before the threat of overflow. The management task is to create enough space in the reservoir to contain the inflows without spilling water. Manoeuvrability implies the ability to run down the reservoir level, and is present only for periods when production can exceed inflow; $\overline{e}_j^H > w_{jt}$. This is the condition for the ability to sustain a *constant* level in the reservoir. Any reservoir level, e.g., the full level, is sustainable within a time period t' to t'' if $\overline{e}_j^H > \max w_{jt}$ for $t \in (t'$ to $t'')$. This is the condition for a potential to prevent overflow at plant j.

If there is a series of high inflow periods spilling may be physically impossible to avoid if emptying the reservoir at the start of the time periods with high inflow and using the maximal production capacity every period, is insufficient to "swallow" all the incoming water. Analogous to the aggregated system case of (3.24) we have an *unavoidable* lock-in situation for plant j when:

$$R_{jt'} = 0, \sum_{t=t'}^{t''} w_{jt} - (t'' - t' + 1)\overline{e}_j^H > \overline{R}_j \ , \tag{4.13}$$

where t' is the start of the high-inflow periods and t'' is the first period with overflow for plant j. Notice that for some periods between t' and t'' the maximal production may be greater than the inflows, but this situation does not remain long enough for the reservoir level to be reduced sufficiently to prevent overflow at t''. This may be the situation for a plant during the period of snow melting or autumn rain illustrated in Figure 1.4 in

Chapter 1. Lock-in situations can occur only at the disaggregated plant level, and for an aggregated system studied in Chapter 3 the aggregation of lock-ins is problematic in the sense that no information relevant for actions is revealed. For management purposes it will be of interest to inspect periods of high inflows (remember that we have assumed perfect knowledge about inflows, i.e., no uncertainty occurs) and to calculate the maximum level of the reservoir preceding the high inflow periods in order to prevent overflow:

$$R_{jt'}^{\max} = \overline{R}_j - (\sum_{t=t'}^{t''} w_{jt} - (t'' - t' + 1)\overline{e}_j^H) \tag{4.14}$$

The lowest possible level of $R_{jt}^{\max}$ is zero [if (4.13) should hold this level would become be negative]. The calculation in (4.14) may also be done for different constellations of the time periods t' and t'' for a fine-tuning of the necessary manoeuvring actions.

Consider we have a development where the situation described in (4.13) holds. Assume that it is actually optimal to have an empty reservoir at the end of period t'. The water values for the time periods in such a series as part of the optimal plan will all be the same from $t'+1$ to t'', and equal to zero, assuming overflow in period t'' only. The water value will become positive again in the period $t''+1$ when the reservoir can be reduced below or to the maximal level since by assumption the inflow is less than the maximal production level in this (and subsequent) periods. This situation is illustrated in Figure 3.8 in Chapter 3 for the aggregated system.

The programming model assigns the extreme value of zero to the shadow price on stored water during the periods from t' to t'', while the output is actually sold to the positive prices of the periods. From the model point of view this is logical, because the accumulation of water ends up with overflow and zero value is assigned to this flow. A marginal increase in accumulation of water has zero value since the reservoir cannot become more than full. A zero water value is just a "go" signal for using as much water as possible from this plant. From a practical point of view the plant creates value in every period of manoeuvring producing at maximal output rate evaluated at the going price. According to (4.12) the shadow price on the production constraint is equal to the social price for each period. A marginal increase in the constraint is evaluated to the current social price. The distinction between shadow value of water as reservoir and shadow value of water being processed is made quite clear.

The example above indicates that there is a potential problem with Hveding's conjecture when the manoeuvrability is not maximal for all plants. Using the test (4.13) above one point is that we may have one reservoir overflowing in period $T-1$ because it is *unavoidable* due to circumstances described by (4.13); there is a lock-in. Otherwise optimal system management will try to avoid a single reservoir overflowing before the others, but the plant-specific manoeuvrability indices are no longer uniformly 1, and the distribution of the manoeuvrability index, coupled with the distribution of plant production constraints, may block the possibility of all plants reaching full reservoirs at the same time. The same reasons hold for emptying reservoirs at the same time being an optimal policy. If (4.13) holds then it may be optimal to empty a reservoir before other reservoirs are emptied in order to minimise the spilling.

If spilling can be avoided, i.e., the situation (4.13) above is not valid, then running one or more periods at maximal output may suffice to avoid overflow. The exact timing of such full production periods will be determined by the overriding objective of maximising consumer plus producer surplus. A decreasing (increasing) price toward the critical overflow period will tend to start early (late) with the manoeuvring, as well as increasing (decreasing) inflows. But the fact that overflow may be avoided may not be the same as to say that Hveding's conjecture holds. It may be that overflow is prevented by some reservoirs being emptied before the others, e.g., manoeuvring is done to accommodate a peak inflow situation when the snow melts. The new crucial aspect of production constraints is that water values may become plant-specific. To treat the system as an aggregated system as the Hveding conjecture invites will then create inaccuracies and lead to loss of objective-function value. But for a group of plants with more or less equal production and reservoir characteristics never experiencing individual water values it will still be the case that Hveding's conjecture is a good approximation to optimal management.

It is obvious that unregulated hydropower and windmills are not covered by Hveding's conjecture since these plants by definition have no storage possibilities. The manoeuvrability index value for unregulated stations is by definition zero. Introducing upper constraints on the rate of production from run-of-the-river plants we will have overflow when inflows exceed these limits:

$$e_{jt}^R \le \overline{e}_j^R ,\ s_{jt}^R = w_{jt}^R - \overline{e}_j^R \ge 0 \text{ for } w_{jt}^R \ge \overline{e}_j^R ,\ j=1,..,N^R , \tag{4.15}$$

where s_{jt}^R is the overflow at unregulated power station j belonging to the group of N^R unregulated power stations. This may be another reason for observing "losing" water while the price is positive.

Windmills also have restrictions on the maximal kW that can be generated. If the wind blows harder than a limit set to protect the survival of the structure with gearbox etc., the windmill has to be turned off and the rotors placed in a neutral position. (This may not help; there are examples of windmills blowing down during extreme winds.)

The existence of such run-by-the-river hydropower plants and also windmills may cause extra adjustment problems for the regulated plants trying to accommodate must-take-power and thus contributes to cases of deviations from Hveding's conjecture without consequences for the objective function value unless lock-in occurs.

Hydraulically coupled hydropower

For hydropower stations located along the same river, the release from upstream reservoirs ends up as inflows to downstream stations.[1] This kind of coupling naturally reduces manoeuvrability of the system. One extreme situation is that downstream dams are fed only by upstream releases. The time lags involved in the couplings depend on the length of the time period. Choosing, e.g., one hour as a time period creates a lag structure of many periods being involved, while choosing a month may result in no lags at all. Focussing on hydraulically coupled stations only and assuming no lag (lags can straightforwardly introduced), the water balance equation for plant j may bc written:

$$\begin{aligned} &R_{jt} \leq R_{j,t-1} + e^{H}_{j-1,t} - e^{H}_{jt}, \\ &\text{where } e^{H}_{j-1,t} = w_{j,t},\ j = 1,..,N^{C},\ t = 1,..,T \end{aligned} \tag{4.16}$$

The hydropower stations are sorted in ascending order going downstream, i.e., j = 0 is the most upstream station, N^C the last station downstream, and (N^C +1) is the number of coupled stations, including the most upstream one. The reservoir accumulation equation for the first plant on the river is $R_{ot} \leq R_{o,t-1} + w_{ot}^{H} - e_{ot}^{H}$. The assumption of no waste of water at the generation stage is maintained, making release equal to production. The inflow to plant j originates as a release at plant j – 1. It is straightforward to include additional current inflows independent of release from an upstream reservoir.

Introducing a group of coupled stations to the model (4.8) for independent

[1] The situation is treated in Wood and Wollenberg (1984), but not in the detail attempted here.

plants the planning problem reads:

$$\max \sum_{t=1}^{T} \int_{z=0}^{x_t} p_t(z)\,dz$$

subject to

$$
\begin{aligned}
& x_t = \sum_{j=0}^{N^C} e_{jt}^H + \sum_{i=1}^{N^I} e_{it}^H ,\ t = 1,..,T \\
& R_{0t} \le R_{0,t-1} + w_{0t}^H - e_{0t}^H \\
& R_{jt} \le R_{j,t-1} + e_{j-1,t}^H - e_{jt}^H ,\ j = 1,..,N^C \\
& R_{it} \le R_{i,t-1} + w_{it} - e_{it}^H ,\ i = 1,..,N^I \\
& R_{jt} \le \overline{R}_j \\
& e_{jt}^H \le \overline{e}_j^H \\
& R_{jt}, x_t, e_{jt}^H \ge 0 \\
& T, w_{jt}, R_{j0}, \overline{R}_j, \overline{e}_j^H \text{ given, } R_{jT} \text{ free, } j = 0,1,..,N,\ t = 1,..,T
\end{aligned}
\tag{4.17}
$$

The number of independent stations is now N^I and coupled stations $N^C + 1$, adding up to N plants. Power from the coupled stations is included in the energy balance. For convenience the index j is used also when pointing to a plant within the total group of plants in the last restrictions in common for both groups (dealing with the first plant on the river then requires special attention).

The Lagrangian function for the problem is, substituting for total consumption and setting $e_{j-1,t}^H = 0$ for $j = 0$:

$$
\begin{aligned}
L = & \sum_{t=1}^{T} \int_{z=0}^{\sum_{j=0}^{N^C} e_{jt}^H + \sum_{i=1}^{N^I} e_{jt}^H} p_t(z)\,dz \\
& - \sum_{t=1}^{T} \sum_{j=0}^{N^C} \lambda_{jt} (R_{jt} - R_{j,t-1} - e_{j-1,t}^H + e_{jt}^H) \\
& - \sum_{t=1}^{T} \sum_{i=1}^{N^I} \lambda_{it} (R_{it} - R_{i,t-1} - w_{it} + e_{it}^H) \\
& - \sum_{t=1}^{T} \sum_{j=1}^{N} \gamma_{jt} (R_{jt} - \overline{R}_j)
\end{aligned}
\tag{4.18}
$$

$$-\sum_{t=1}^{T}\sum_{j=1}^{N}\rho_{jt}(e_{jt}^{H}-\bar{e}_{j}^{H})$$

We are only interested in the necessary first-order conditions for the coupled stations since the conditions for independent plants have been dealt with previously:

$$\begin{aligned}
&\frac{\partial L}{\partial e_{jt}^{H}}=p_t(\sum_{j=0}^{N^C}e_{jt}^{H}+\sum_{i=1}^{N^I}e_{it}^{H})-\lambda_{jt}+\lambda_{j+1,t}-\rho_{jt}\le 0\ (=0 \text{ for } e_{jt}^{H}>0)\\
&\frac{\partial L}{\partial R_{jt}}=-\lambda_{jt}+\lambda_{j,t+1}-\gamma_{jt}\le 0\ (=0 \text{ for } R_{jt}>0)\\
&\lambda_{jt}\ge 0(=0 \text{ for } R_{jt}<R_{j,t-1}+w_{jt}-e_{jt}^{H})\\
&\gamma_{jt}\ge 0(=0 \text{ for } R_{jt}<\bar{R}_{j})\\
&\rho_{jt}\ge 0(=0 \text{ for } e_{jt}^{H}<\bar{e}_{j}^{H})\ , \qquad t=1,..,T\,, j=0,1,..,N^C
\end{aligned} \tag{4.19}$$

The conditions for the independent stations remain the same as in (4.12). Notice that outputs from all independent plants also are added to form total supply in the demand function. There is only a need to consider two consecutive plants downstream at a time. The first condition in (4.19) shows the only change in the first-order conditions for coupled plants: the water value for the next downstream plant is *added* to the social price showing the value of processing water at plant j for time period t. Having the released water utilised one more time *increases* the water value of the upstream plant relative to the current social price. Assuming that all coupled plants are producing and that we have interior solutions for all of them implies that the water value for the last plant, N^C, is equal to the social price for the period in question:

$$p_t(\sum_{j=0}^{N^C}e_{jt}^{H}+\sum_{i=1}^{N^I}e_{it}^{H})=\lambda_{N^C t} \tag{4.20}$$

For the next plant N^C-1 upstream the water value is:

$$\begin{aligned}
&p_t(\sum_{j=0}^{N^C}e_{jt}^{H}+\sum_{i=1}^{N^I}e_{it}^{H})=\lambda_{N^C-1,t}-\lambda_{N^C t}\Rightarrow\\
&\lambda_{N^C-1,t}=2p_t(\sum_{j=0}^{N^C}e_{jt}^{H}+\sum_{i=1}^{N^I}e_{it}^{H})
\end{aligned} \tag{4.21}$$

For plant j the water value then becomes:

$$\lambda_{jt} = (N^C + 1 - j)p_t(\sum_{j=0}^{N^C} e_{jt}^H + \sum_{i=1}^{N^I} e_{it}^H) \tag{4.22}$$

The water values upstream become greater than the social price because the same water can be utilised several times downstream. If there are time lags involved in the appearance of inflows downstream, the appropriate dating of the social prices will be reflected in the sum over prices in (4.22).

The question is how this situation will influence the optimal utilisation of the reservoirs. If upstream releases are the only inflows downstream, pure accumulation of water from the first period in the first plant means that production also stops at all down-stream plants assuming for convenience that the initial levels of the reservoirs are all zero. The water value at plant $j = 0$ must be higher (or equal to) $N^C p_t$ for this to be optimal. The first reaction may be that this is highly improbable and the plant should never accumulate all inflows. But the situation of the coupled plant is really not different than for an independent plant, because any future social price will be inflated with the same factor when forming the water value. The shadow prices of coupled plants are not independent of the social price, but expressed as multiples. The shadow-price dynamics will be the same as for independent plants. The general storage philosophy will be the same for coupled plants; maximal water should be transferred to high-price periods. If there is not enough water to go around the first reservoir should be full and then in the natural priority order downstream.

The management problems with hydraulic couplings are the difficulties posed for the manoeuvring of the system as regards keeping within the reservoir constraints when there are production restrictions. A downstream plant not only has to know the release of the next upstream plant to determine its own release in order to avoid spilling, but spilling may be unavoidable if the downstream plant hits its production constraint. In the case of production constraints along the river this task becomes quite involved. Coupled plants can therefore not be treated as independent plants when working out an optimal management plan. This has a direct implication for the possibility of realising a social optimal plan using a decentralised market.

Environmental restrictions

As pointed out in the comments to Table 3.1 of the constraint taxonomy for hydropower plants, there may be constraints on both maximal and minimal releases to a continuing watercourse due to considerations of down-stream activities. One such activity is just another hydropower plant downstream as expressed by (4.16). Now, maintaining the assumption of no waste of water at the production stage of electricity, production can be substituted for release of water. The model (4.8) then already covers the upper constraint. The only change we may want to make is to introduce a period-dependent upper level as shown in Table 3.1 for water release.

Restrictions on ramping up and down may be most relevant for hydraulically coupled plants, but for convenience we revert to model (4.8) with independent plants only. A combination of different types of plants is straightforward. Substituting actual production for releases yields the following constraints concerning releases and ramping:

$$\begin{aligned} &0 \le \underline{e}_{jt}^{H} \le e_{jt}^{H} \le \overline{e}_{jt}^{H}, \\ &e_{jt}^{H} - e_{j,t-1}^{H} \le e_{jt}^{ru}, \\ &e_{j,t-1}^{H} - e_{jt}^{H} \le e_{jt}^{rd}, t = 1,..,T, j = 1,..,N \end{aligned} \tag{4.23}$$

The total release and ramping-up and -down restrictions for period t are expressed by $\overline{e}_{jt}^{H}, \underline{e}_{jt}^{H}, e_{jt}^{ru}, e_{jt}^{rd}$ respectively, in accordance with the expressions in Table 3.1 in Chapter 3. These restrictions depend on time, since environmental impacts may vary with both period of the day and season. A production constraint independent of time as in (4.8) is not specified for ease. The planning problem becomes:

$$\begin{aligned} &\max \sum_{t=1}^{T} \int_{z=0}^{x_t} p_t(z)\,dz \\ &\text{subject to} \\ &x_t = \sum_{j=1}^{N} e_{jt}^{H}, t = 1,..,T \\ &R_{jt} \le R_{j,t-1} + w_{jt} - e_{jt}^{H} \\ &R_{jt} \le \overline{R}_{j} \\ &0 \le \underline{e}_{jt}^{H} \le e_{jt}^{H} \le \overline{e}_{jt}^{H}, \\ &0 \le e_{jt}^{H} - e_{j,t-1}^{H} \le e_{jt}^{ru}, \\ &0 \le e_{j,t-1}^{H} - e_{jt}^{H} \le e_{jt}^{rd} \\ &R_{jt}, e_{jt}^{H} \ge 0 \end{aligned} \tag{4.24}$$

$$T, w_{jt}, R_{jo}, \overline{R}_j, \underline{e}_{jt}^H, \overline{e}_{jt}^H, e_{jt}^{ru}, e_{jt}^{rd} \text{ given}, R_{jT} \text{ free},$$
$$j = 1,..,N,\ t = 1,..,T$$

Substituting for total consumption from the energy balance the corresponding Lagrangian function is:

$$\begin{aligned}
L = & \sum_{t=1}^{T} \int_{z=0}^{\sum_{j=1}^{N} e_{jt}^H} p_t(z)dz \\
& -\sum_{t=1}^{T}\sum_{j=1}^{N} \lambda_{jt}(R_{jt} - R_{j,t-1} - w_{jt} + e_{jt}^H) \\
& -\sum_{t=1}^{T}\sum_{j=1}^{N} \gamma_{jt}(R_{jt} - \overline{R}_j) \\
& -\sum_{t=1}^{T}\sum_{j=1}^{N} \overline{\rho}_{jt}(e_{jt}^H - \overline{e}_{jt}^H) \\
& -\sum_{t=1}^{T}\sum_{j=1}^{N} \underline{\rho}_{jt}(-e_{jt}^H + \underline{e}_{jt}^H) \\
& -\sum_{t=1}^{T}\sum_{j=1}^{N} \psi_{jt}^{ru}(e_{jt}^H - e_{j,t-1}^H - e_{jt}^{ru}) \\
& -\sum_{t=1}^{T}\sum_{j=1}^{N} \psi_{jt}^{rd}(e_{j,t-1}^H - e_{jt}^H + e_{jt}^{rd})
\end{aligned} \tag{4.25}$$

The shadow prices for the restriction for releases, and ramping up and down are $\overline{\rho}_{jt}, \underline{\rho}_{jt}, \psi_{jt}^{ru}, \psi_{jt}^{rd}$. When deriving the necessary first-order conditions for period t we must remember that the release of water during period t also appears in the ramping restrictions in period $t + 1$:

$$\begin{aligned}
& \frac{\partial L}{\partial e_{jt}^H} = p_t(\sum_{i=1}^{N} e_{it}^H) - \lambda_{jt} - \overline{\rho}_{jt} + \underline{\rho}_{jt} - \psi_{jt}^{ru} + \psi_{jt}^{rd} + \psi_{j,t+1}^{ru} - \psi_{j,t+1}^{rd} = 0 \\
& \frac{\partial L}{\partial R_{jt}} = -\lambda_{jt} + \lambda_{j,t+1} - \gamma_{jt} \le 0\ (= 0 \text{ for } R_{jt} > 0) \\
& \lambda_{jt} \ge 0 (= 0 \text{ for } R_{jt} < R_{j,t-1} + w_{jt} - e_{jt}^H) \\
& \gamma_{jt} \ge 0 (= 0 \text{ for } R_{jt} < \overline{R}_j) \\
& \overline{\rho}_{jt} \ge 0 (= 0 \text{ for } e_{jt}^H < \overline{e}_{jt}^H)
\end{aligned} \tag{4.26}$$

$$\underline{\rho}_{jt} \geq 0 (= 0 \text{ for } e^H_{jt} > \underline{e}^H_{jt})$$
$$\psi^{ru}_{jt} \geq 0 (= 0 \text{ for } e^H_{jt} - e^H_{j,t-1} < e^{ru}_{jt})$$
$$\psi^{rd}_{jt} \geq 0 (= 0 \text{ for } e^H_{j,t-1} - e^H_{jt} < e^{rd}_{jt}) \,, \qquad t = 1,..,T \,, j = 1,..,N$$

The shadow prices for the release and ramping constraints show up in the first condition for the optimal adjustment of production for unit j for period t. The condition must hold with equality since water release is constrained to be positive. Upper and lower production and ramping constraints cannot both be binding at the same time, so in the first condition in (4.26) not more than three of the shadow prices concerning total release and ramping can be positive at the same time.

However, we should observe the connection between production and ramping constraints. Combining the ramping-up constraint and the upper-production constraint we have:

$$e^H_{jt} \leq e^{ru}_{jt} + e^H_{j,t-1}, \; e^H_{jt} \leq \bar{e}^H_{jt} \tag{4.27}$$

This means that only one of the constraints can become binding, determined by which of the expressions $(e^{ru}_{jt} + e^H_{j,t-1})$ and $\bar{e}^H_{jt}$ is the greatest. In a similar way, combining the ramping-down constraint and the lower production constraint we have:

$$e^H_{jt} \geq e^H_{j,t-1} - e^{rd}_{jt}, \; e^H_{jt} \geq \underline{e}^H_{jt} \tag{4.28}$$

Again, only one of the constraints can become binding, determined by which of the expressions $(e^H_{j,t-1} - e^{rd}_{jt})$ and $\underline{e}^H_{jt}$ is the greatest.

As expanded upon in the case of an upper production constraint previously, we have a situation with social prices and water values not necessarily coinciding. It is only shadow prices concerning the water in the reservoirs that appear in the dynamic equation in (4.26). The shadow prices on the environmental constraints do not enter the dynamic equation, but influences the price formation through interactions with the demand side.

If the lower-release restriction that is new compared with conditions (4.12), is binding, but no other constraint, then we have that the water value for plant j will potentially be higher than the social price for the same period. More water is processed than what would be optimal without the restriction. In addition to the social price, the water value may reflect a premium for fulfilling an environmental constraint. To see whether it is feasible to have water value higher than the price as part of an optimal plan must be checked. There are three social price regimes to investigate for two time periods:

i) $p_t = p_{t+1}$
ii) $p_t > p_{t+1}$
iii) $p_t < p_{t+1}$

Assume first that the prices for the periods t and $t + 1$ are the same. According to the dynamic shadow-price condition in (4.26) the water values become the same, provided that the shadow price on the reservoir constraint in period t is zero. This will be the case if there is no threat of overflow in period t. Assuming that the minimum-flow condition is not binding in period $t + 1$, it is not possible for the water value in period t to be higher than the price, i.e., the shadow price on the minimum drawing of water is zero. It is not logical to have a threat of overflow in period t since it is the minimum water-use constraint that is binding. This implies that decreasing the minimum water constraint for plant j will not influence the value of the objective function in the optimal management plan.

Now assume that the price in period t is higher than the price in period $t + 1$, maintaining that the minimum water constraint is not binding in the latter period. Then the water value in period t should be *lower* than the price in period t, which is a contradiction of the assumption. Such a constellation of prices must then be ruled out.

The last case of a lower price in period t than $t + 1$ is the case consistent with how forced use of water may interact with demand in the price formation. The water value in period $t + 1$ is by assumption of no binding environmental constraint in the period equal to the social price, which again is equal to the water value in period t via the reservoir-related shadow-price dynamics in (4.26). The water value is then greater than the price in period t, allowing for a positive shadow price of the minimum-water constraint in the period so the first condition in (4.26) can be fulfilled with equality. The social evaluation of production in period t is lower than the water value because the reference for the water value is the value the stored water in period t can create when used in period $t + 1$. The positive value of the shadow price on the minimum-water constraint is not dictated by the minimum water-flow constraint as such, but by the difference between the current price and the price prevailing when more water than the minimum amount is processed. The difference between the price and the water value is not a reward for processing a minimum amount of water, but expresses the extra value of the water reaped if waiting with processing it to a later period when the social price will be higher.

The situation that the price in period t is lower than the price in period $t + 1$ is the typical case for accumulating water to be used in higher-price periods, especially for multiyear reservoirs as mentioned previously in the chapter. In the pure accumulation case the water values during intervals with no production became equal to the social price in the first period resuming production. Now we have production in all periods due to the minimum water-flow requirement, but this fact does not influence the water-value dynamics. The water values during periods with keeping the minimum production become equal to the water value, i.e., the social price, in the first period when the minimum water use is exceeded, assuming the reservoir not to be full in this period, and the minimum-water constraint not being binding. A minimum water-flow will slow the accumulation of water in plants with multiyear reservoir capacities.

The water value in period t may in general be higher than the price in period t if there is no threat of overflow in period t, which is quite logical if the minimum water-flow constraint is binding.

Concerning ramping constraints the discussion of shadow prices with negative signs in the first condition in (4.26) will follow the discussion of the shadow price on the upper production constraint, and discussion of shadow prices with positive signs will follow the discussion of the shadow price on the lower production constraint. Ramping constraints are, of course, not relevant for plants keeping constant production. If it is assumed that production constraints dominate according to the relevant condition contained in (4.27) and (4.28), ramping constraints for period t are superfluous, or if the ramping constraints dominate the discussion of production constraints is superfluous. However, a unique feature is that ramping constraints for the next period $t + 1$ enters the decision about production today. This interconnectedness of production levels and ramping constraints in different periods complicate the simultaneous solution to the dynamic multi-period planning problem.

Concerning Hveding's conjecture when both upper and lower production constraints and/or ramping constraints are present, the conjecture may still work for a subgroup of plants with "weak enough" constraints, but the more constraints there are the more the manoeuvrability is reduced and the greater possibility for locking in of water, and creating plant-specific water values, making the conjecture invalid. Simple summation of reservoirs and upper capacities may become too misleading in the face of such environmental constraints as introduced above.

Chapter 5. Mix of Thermal and Hydropower Plants

Norway is unique in having almost only hydropower plants generating all the electricity. But other countries that rely to a high degree on hydro must have other forms of generating plants in a mix that varies from country to country. Norway participates in the international wholesale electricity market Nord Pool together with Denmark, Finland, and Sweden, where in 2003 the hydro share was 46%, conventional thermal was 28%, nuclear power 24% (increasing when a new Finnish station is planned to come on stream in 2010), and wind power 2%. It is therefore of interest to include other forms of generation and to study how the running of such capacities interacts with the operation of hydropower plants. We will focus on the class of generators termed *thermal* plants. As mentioned in Chapter 1, the operational problem of hydropower plants with reservoirs is essentially a dynamic problem, while the running of thermal plants will mainly be a static problem. Hydro plants are usually energy constrained, while thermal plants are effect-constrained. Thus the interaction may be of a special type.

The load curve for a yearly period is illustrated in Figure 1.3 in Chapter 1. With a mix of plants attention can be paid to the load curve when constructing thermal capacity. The design and choice of technology and scale can influence the relationship between fixed costs, overwhelmingly consisting of capital cost, and variable costs. It may be part of a cost-efficient choice, considering both investment and operating costs, to use capacity with low investment cost per kW of total capacity, but with higher variable costs for peak periods. Similarly, capacity with low variable cost but higher investment costs may be cost-efficient as base load. The role as to peak load or base load use taken by various forms of generating capacities will be of special interest when hydro is involved.

Thermal plants

Thermal plants use fossil fuels as energy source, like coal, oil, gas, and wood, either to heat up water and using steam to run turbines, or directly such as combustion technologies developed for gas. In industries using steam for production purposes, like the pulp and paper industry, the steam may be used also to generate electricity as a joint product. There are other forms of co-generation, like at district heating plants. It is usual to include nuclear power plants among thermal plants. The heat created by the reactor is used to make steam that drives the turbines.

The environmental problems created by running thermal plants are widespread and serious, both on a regional scale and a global scale. Acid rain causes damage to vegetation of various types from forests to crops, to aquatic life, especially fish populations, corrosion on surfaces of buildings and respiratory health problems. The active components in the emissions are sulphur, nitrogen, and particles, all stemming from the primary energy input mainly through combustion. Global warming problems are created by emissions of carbon dioxide. Nuclear power plants create insignificant emissions in normal running. The problems are long-run ones of creation of nuclear waste and the probability of operational accidents. Although the probability may be extremely low, the damage may also be extremely high as we saw after the Chernobyl accident.

The short-run production function for thermal plants (may be exclusive of nuclear plants) may in a simple way be expressed by:

$$\begin{aligned} e_{it}^{Th} &= f_{it}(E_{is,t}, L_{it}),\ \frac{\partial f_{it}}{\partial E_{is,t}} > 0, \frac{\partial f_{it}}{\partial L_{it}} \geq 0,\ i = 1,.., M,\ s = 1,.., S \\ z_{ip,t} &= g_{ip,t}(E_{is,t}, L_{it}),\ \frac{\partial g_{ip,t}}{\partial E_{is,t}} > 0, \frac{\partial g_{ip,t}}{\partial L_{it}} \leq 0,\ p = 1,.., P \end{aligned} \tag{5.1}$$

Here e_{it}^{Th} is production of electricity from thermal plant i, of a total of M plants, using primary energy input vector $E_{is,t}$, and labour input L_{it}. Both inputs have positive marginal productivities, but the labour input may have a zero marginal productivity impact if labour has the role of overseeing processes rather than doing activities directly related to the rate of production. The energy input indexed s may often be a single primary energy like coal, etc., or it may be a vector of several types at the same time. The emission vector, $z_{ip,t}$, is created as a by-product of the production of electricity, and is a function of the same inputs as electricity. The pollutant index p may run over sulphur, nitrogen, particles, etc. The uses of the primary energy inputs give rise to one or several pollutants. These forms of production relations are termed *factorially determined multi-output production* in Frisch (1965).

Capital is not shown as a factor of production in (5.1), but incorporated in the functional form since capital is given in the short run. We do not bother to introduce capacity limits on production here, but return to this when introducing the corresponding cost function. The technologies may depend on time, as indicated by the time subscripts on the production functions. If technology is disembodied technical change may occur smoothly over time. In the case of embodied technical change investments are needed to influence the technology of the short-run functions (Johansen, 1972; Førsund and Hjalmarsson, 1987).

Abatement possibilities are not specified explicitly, but we may expand labour to be of two categories, production workers and abatement workers, and thus model abatement, assuming $\partial g_{ip,t}/\partial L_{it} < 0$ for abatement workers and zero marginal productivity in the $g(.)$ function for production workers, and correspondingly, in the production function for electricity the marginal productivity for abatement workers is zero, while it is positive for production workers. The choice of a production technology $f(.)$ may dictate the emission technology $g(.)$, e.g., in such a way that a more expensive technology to run implies an emission technology generating less emissions for a given amount of primary energy, thus the choice of technology is also an abatement decision.

To serve our purpose of studying the interaction with hydro, the other forms of generation are not studied in detail. Furthermore, we will not pursue the emission theme, but just point out how the emissions can be taken into consideration generating electricity from different types of generators.

We will use short-run variable cost functions as functions of electricity output with given explicit capacity limits in the short run. The cost function is derived in the standard way of minimising outlays on variable inputs for a given level of electricity output, and subject to environmental regulation:

$$\begin{aligned}
&\min \left(\sum_{s=1}^{S} q_{is,t} E_{is,t} + \omega_{it} L_{it}\right) \\
&\text{subject to} \\
&e_{it}^{Th} = f_{it}(E_{is,t}, L_{it}),\ e_{it}^{Th} \text{ given, } i = 1,..,M,\ s = 1,..,S \\
&z_{ip,t} = g_{ip,t}(E_{is,t}, L_{it}) \le \overline{z}_{ip,t} \text{ given, } \ p = 1,..,P\ ,
\end{aligned} \tag{5.2}$$

where $q_{is,t}$ is the price of primary energy input s and ω_{it} the labour cost. Environmental aspects may be taken care of by imposing an upper level of emissions $\overline{z}_{ip,t}$ from each plant as done in (5.2), and/or by introducing technology standards (not shown explicitly). The standard assumptions from economic cost analysis will be entertained, although a more detailed insight may reveal deviations from textbook assumptions of smooth convex functions. A special feature of start-up costs will be discussed briefly later. What is of especial interest is that start-up, and also closedown costs, make the detailed running of thermal plants a dynamic problem.

Building on the solution of problem (5.2), a plant-specific variable cost function for the generation of electricity based on thermal energy sources is introduced. Each plant has an upper capacity, $\overline{e}_{it}^{Th}$, for generation that can be changed only by investments. The concept of a given capacity is not necessarily uniquely defined in practice, but here it will mean the capacity at a normal operating situation of the station, i.e., it may be possible to squeeze more out of the plant in the short run, but up to a level that is not sustainable without breakdowns in a longer perspective.

For simplicity the cost functions are not dated, but the cost function may in the real world change between periods in the relatively short run due to changing primary fuel prices. Fuels may be more expensive in a high-demand season, or be subject to a price drift over time, and technology may also change over time due to technical change, or due to a change in environmental policy, e.g., changing the upper levels on emissions specified in (5.2). For simplicity we keep factor prices and technology constant:

$$c_{it} = c_i(e_{it}^{Th}),\ c_i' > 0, c_i'' > 0,\ e_{it}^{Th} \le \overline{e}_i^{Th},\ i = 1,..,M, t = 1,..,T \tag{5.3}$$

In contrast to these standard economic textbook assumptions, a plant may be designed to have the smallest average, and maybe also marginal, cost at close to full capacity utilisation; i.e., marginal cost, as well as average variable cost, is decreasing up to normal capacity. This shape of the variable cost function may explain why a conventional thermal unit may be closed down when its capacity utilisation rate drops below 40% as is often stated by engineers. Such possibilities are disregarded here and a standard assumption of increasing marginal cost will be entertained. We disregard, for the time being, also costs of ramping up or down plants, and especially going from a cold to a spinning state. A plant in a spinning state is producing below the capacity, maybe down to zero, but the production can increase fairly fast.

The case of linear, but different cost functions is illustrated in Figure 5.1. The arrows marked 1, 2, and 3 represent three total variable cost functions with different marginal costs. The base-load cost function 1 is the cheapest to run per unit of output, then comes the shoulder capacity 2 and last the most expensive peak-load capacity 3. (Remember that investment costs per unit of maximal capacity $\overline{e}_i^{Th}$ are not shown.) The capacity limits of the three technologies are indicated on the horizontal axis. Running each activity in a cost-efficient way results in the region of possible cost output combinations for the three units shown by the faceted "diamond"

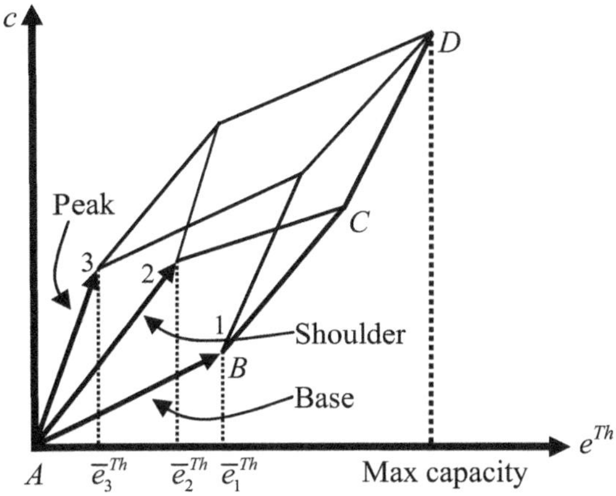

Figure 5.1. Linear total variable cost functions.

ABCDA going counter-clockwise. Obviously the curve *ABCD* describes the least-cost way of using the capacities, and the maximum output is defined as $\sum_{i=1}^{M} \bar{e}_i^{Th}$.

The least cost combination of thermal plants, satisfying a total generating requirement of e_t^{Th} for each period, is found by solving the following problem:

$$\begin{aligned} &\min \sum_{i=1}^{M} c_i(e_{it}^{Th}) \\ &\text{subject to} \\ &\sum_{i=1}^{M} e_{it}^{Th} \geq e_t^{Th} \\ &e_{it}^{Th} \leq \bar{e}_i^{Th} \\ &e_{it}^{Th} \geq 0 \\ &e_t^{Th}, \bar{e}_i^{Th} \text{ given, } t = 1,..,T,\ i = 1,..,M \end{aligned} \tag{5.4}$$

The corresponding Lagrangian function (converting the problem to one of maximisation to ease the comparison with the set-up in Sydsæter et al., 2005), is

$$L = -\sum_{i=1}^{M} c_i(e_{it}^{Th})$$

$$-\nu(-\sum_{i=1}^{M} e_{it}^{Th} + e_{t}^{Th})$$
$$-\sum_{i=1}^{M} \theta_i (e_{it}^{Th} - \overline{e}_{i}^{Th}) \tag{5.5}$$

The necessary first-order conditions are:

$$\frac{\partial L}{\partial e_{it}^{Th}} = -c_i'(e_{it}^{Th}) + \nu - \theta_i \leq 0 \ \ (= 0 \text{ for } e_{it}^{Th} > 0)$$
$$\nu \geq 0 (= 0 \text{ for } \sum_{i=1}^{M} e_{it}^{Th} > e_{t}^{Th}) \tag{5.6}$$
$$\theta_i \geq 0 (= 0 \text{ for } e_{it}^{Th} < \overline{e}_{i}^{Th})$$

A concave objective function and convex constraints in (5.5) are sufficient conditions for a maximum. Notice that the more realistic functional forms for short-run cost functions mentioned above would violate the concavity of the objective function. (However, in the case of falling marginal cost curves there may be a unique solution running all but one plant at maximal capacity if the marginal cost curves do not intersect.) A plant will not be used in period t if the marginal cost is greater than the shadow price on the total production requirement. Since it is not used, the shadow price on its capacity constraint is zero, according to the last complementary-slackness condition in (5.6). Plants in use will face the same marginal costs as long as the shadow prices on the capacity constraints remain zero. At total production requirement, just exhausting the capacity of a plant, the shadow price on the capacity constraint typically becomes positive. The marginal cost of this plant is then the difference between the shadow price on the total production requirement constraint and the shadow price on the capacity constraint. A marginal increase in the production requirement necessitates the use of a more expensive unit, while an expansion of the capacity of the constraining unit keep us at this unit's level of marginal cost. For plants in use rearranging the first condition in (5.6) yields:

$$c_{i'}'(\overline{e}_{i'}^{Th}) + \theta_{i'} = c_{i''}'(e_{i''t}^{Th}) = \nu \tag{5.7}$$

where the index i' belongs to fully utilised plants and i'' to partially utilised plants. For each level of total generation we get a set of plants producing positive output and a set being idle, according to the marginal cost levels. If the range of variation in the marginal costs for each plant is sufficiently small so that no interval is overlapping, all but one plant will be utilised to full capacity, and there will be a single marginal unit partially utilised. The most expensive plant in use will then as a rule produce below the capacity limit, while all other plants in use are fully utilised. A *merit-order ranking* in this situation means that the cost function for the thermal sector can be arranged starting with the unit with the lowest marginal cost (i.e., highest shadow price on the capacity constraint) at full capacity up to the marginal unit.

This situation may be illustrated as in Figure 5.2 based on the linear total variable cost functions portrayed in Figure 5.1. The marginal costs functions are straight lines, and the locations of base, shoulder and peak load are indicated by the numbering 1, 2, and 3. The capacities are indicated on the horizontal axis. In addition to the individual short-run marginal cost curves, a step-curve denoted *AA'BB'CD* is shown, corresponding to the total piecewise linear cost curve *ABCD* in Figure 5.1. This is the supply curve of the thermal sector. Two levels of total production are shown, one level coinciding with the capacity limit of plant 1, and a second level indicated by the vertical broken line on the horizontal axis. In the first situation the production capacity of plant 1 is just exhausted, so the marginal cost of plant 1 is equal to the difference between the shadow price on the production requirement constraint and the capacity constraint. In the second situation the marginal cost of plant 3 is equal to the shadow price on the production requirement constraint.

We can perform a merit order ranking of the active units according to average variable costs at full capacity utilisation. This ranking will correspond to the general supply curve of the thermal sector in the situation of variable and falling marginal costs up to the capacity limit under the as-

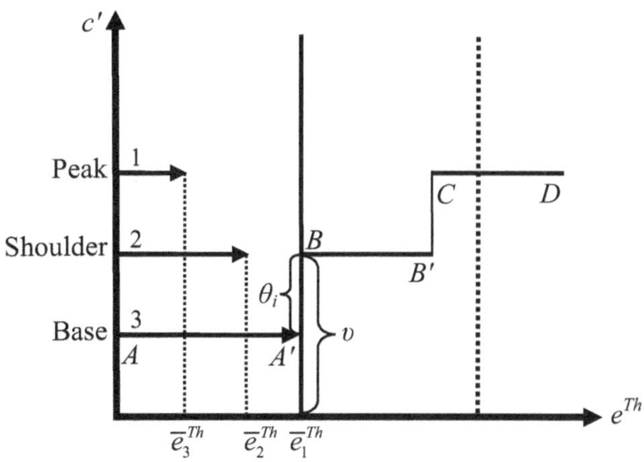

Figure 5.2. Marginal costs and merit order.

sumption that the intervals for the marginal costs curves do not overlap. In other more complex cases the supply curve may be unique to each total output requirement in the sense that a merit order ranking may change the set and order of plants from one total level to another. In the situation of linear variable cost curves, applying the optimal conditions leads to the total variable cost curve *ABCD* being the least cost solution to problem (5.4).

If a merit-order ranking of individual thermal plants is unique we may then aggregate over individual plants by using this ranking as the sector's supply curve. It may formally be approximated by postulating a relationship between total output and total costs:

$$c_t = c(e_t^{Th}), c' > 0, c'' > 0,\ e_t^{Th} \le \sum_{i=1}^{N} \overline{e}_{it}^{Th} \equiv \overline{e}^{Th} \tag{5.8}$$

In the case of linear variable cost functions the sequence of individual cost curves can be simplified or approximated by a smooth function represented by (5.8). In order to represent the realistic situation that the marginal costs of the least expensive plant is positive, as in Figures 5.1 and 5.2, we will assume that $c'(0) > 0$. The various types of capacities may then be defined by delimitating relevant parts of the marginal cost curve, as illustrated in Figure 5.3 based on smoothing the step-curve in Figure 5.2 by fitting a marginal cost curve $c'c'$.

The merit-order ranking leading to the aggregate supply curve for thermal capacity may be regarded as the analogue to Hveding's conjecture for aggregating individual hydropower plants.

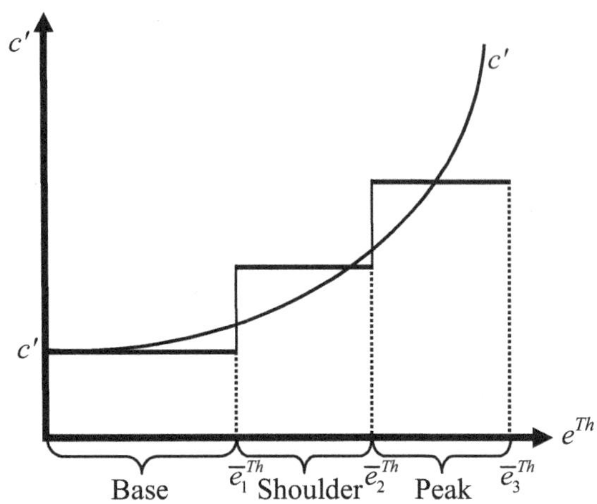

Figure 5.3. Aggregation of marginal cost curves.

The emissions from thermal generation are expressed in (5.2). Environmental policy may influence the merit-order ranking. One way of seeing this is to introduce emissions in problem (5.4). In order to simplify we only look at one type of emission and connect its level to the output level at a plant. The least-cost combination of plants with a total emission constraint is then found by solving the following problem:

$$\begin{aligned}
&\min \sum_{i=1}^{M} c_i(e_{it}^{Th}) \\
&\text{subject to} \\
&\sum_{i=1}^{M} e_{it}^{Th} \geq e_t^{Th} \\
&e_{it}^{Th} \leq \bar{e}_i^{Th} \\
&\sum_{i=1}^{M} z_{it} \leq \bar{z}_t \\
&e_{it}^{Th}, z_{it} \geq 0, \\
&e_t^{Th}, \bar{e}_i^{Th}, \bar{z}_t \text{ given}, i = 1,..,M, t = 1,..,T
\end{aligned} \tag{5.9}$$

The single type of emission is z_{it}, and an environmental objective, $\bar{z}_t$, is introduced for the sector. The objective may vary with period, e.g., emission constraints being lower in winter time than summer time, or vice versa depending on climatic conditions, occurrence of air inversions, etc. The emission from each plant is connected to the production level by

$$z_{it} = g_{it}(e_{it}^{Th}), g_{it}' > 0, i = 1,..,M \tag{5.10}$$

which represents a simplification of the Frisch (1965) multi-output production model in (5.2).Substituting for emissions using (5.10) the Lagrangian function is:

$$\begin{aligned}
L = &-\sum_{i=1}^{M} c_i(e_{it}^{Th}) \\
&-\nu(-\sum_{i=1}^{M} e_{it}^{Th} + e_t^{Th}) \\
&-\sum_{i=1}^{M} \theta_i(e_{it}^{Th} - \bar{e}_i^{Th}) \\
&-\mu(\sum_{i=1}^{M} g_{it}(e_{it}^{Th}) - \bar{z}_t)
\end{aligned} \tag{5.11}$$

The necessary first-order conditions are:

$$\begin{aligned}
&\frac{\partial L}{\partial e_{it}^{Th}} = -c_i'(e_{it}^{Th}) + \nu - \theta_i - \mu g_{it}' \leq 0 \ \ (=0 \text{ for } e_{it}^{Th} > 0) \\
&\nu \geq 0 (=0 \text{ for } \sum_{i=1}^{M} e_{it}^{Th} > e_t^{Th}) \\
&\theta_i \geq 0 (=0 \text{ for } e_{it}^{Th} < \overline{e}_i^{Th}) \\
&\mu \geq 0 (=0 \text{ for } \sum_{i=1}^{M} g_{it}(e_{it}^{Th}) < \overline{z}_t)
\end{aligned} \tag{5.12}$$

Looking at plant in use the first condition in (5.12) can be written, analogously with (5.7):

$$c_{i'}'(\overline{e}_t^{Th}) + \theta_{i'} + \mu g_{i't}' = c_{i''}'(e_{i''t}^{Th}) + \mu g_{i''t}' = \nu \tag{5.13}$$

Both plants that are fully used and partially used get an additional cost term reflecting the environmental policy, assuming that the environmental constraint is binding with a positive shadow price. This term is dependent on the individual unit, and therefore this term will in general influence the merit-order ranking and may result in a different ranking than the one depending only on production costs.

Social solution of mixed hydro and thermal capacity

In the case of an aggregate hydro sector we introduce thermal capacity modelled by the aggregate variable cost relation (5.8). The basic hydro model (2.4) in Chapter 2 without constraints on reservoirs, but only a constraint on total availability of water, is adopted. However, as stated in Chapter 2, this does not mean that we have to assume all inflows arriving in the first period. We saw in Chapter 3 in the case of specifying the reservoir accumulation dynamics and introducing a reservoir constraint, that if the upper and lower constraints are never reached, then the price will be the same for all periods. This means that under such circumstances we can specify a total water constraint and drop to show the reservoir accumulation equation and upper reservoir constraint. The general objective function (3.1) is used maximising consumer plus producer surplus as in problem (3.3) in Chapter 3. We assume that it does not matter how electricity is generated, i.e., the willingness to pay is the same for the two types of generation (no "green" preferences). The energy balance in consumption x_t

and total production describing the physical electric equilibrium is then:

$$x_t = e_t^H + e_t^{Th},\ t = 1,..,T \tag{5.14}$$

When setting up the consumer plus producer surplus the cost function for the thermal sector must now be deducted when expressing this surplus.

The optimisation problem faced by a system planner is:

$$\begin{aligned}
&\max \sum_{t=1}^{T}[\int_{z=0}^{x_t} p_t(z)dz - c(e_t^{Th})] \\
&\text{subject to} \\
&x_t = e_t^H + e_t^{Th} \\
&\sum_{t=1}^{T} e_t^H \le W \\
&e_t^{Th} \le \overline{e}^{Th} \\
&x_t, e_t^H, e_t^{Th} \ge 0,\ t = 1,..,T \\
&T, W, \overline{e}^{Th} \text{ given}
\end{aligned} \tag{5.15}$$

Inserting the energy balance the Lagrangian function is:

$$\begin{aligned}
L = &\sum_{t=1}^{T}[\int_{z=0}^{e_t^H + e_t^{Th}} p_t(z)dz - c(e_t^{Th})] \\
&-\sum_{t=1}^{T}\theta_t(e_t^{Th} - \overline{e}^{Th}) \\
&-\lambda(\sum_{t=1}^{T} e_t^H - W)
\end{aligned} \tag{5.16}$$

The necessary conditions are:

$$\begin{aligned}
&\frac{\partial L}{\partial e_t^H} = p_t(e_t^H + e_t^{Th}) - \lambda \le 0 \ (= 0 \text{ for } e_t^H > 0) \\
&\frac{\partial L}{\partial e_t^{Th}} = p_t(e_t^H + e_t^{Th}) - c'(e_t^{Th}) - \theta_t \le 0 \ (= 0 \text{ for } e_t^{Th} > 0) \\
&\lambda \ge 0 \ (= 0 \text{ for} \sum_{t=1}^{T} e_t^H < W) \\
&\theta_t \ge 0 \ (= 0 \text{ for } e_t^{Th} < \overline{e}^{Th})
\end{aligned} \tag{5.17}$$

Assuming that electricity must be produced in all periods we must then in each period either activate hydro or thermal, or both. Thermal will not be used for periods when

$$c'(0) > \lambda \tag{5.18}$$

If the marginal cost curve starts at values greater than the water value, then thermal is not used. According to the last complementary slackness condition in (5.12) $\theta_t = 0$ when $e_t^{Th} = 0$.

Combining the conditions in (5.17) hydro will not be used in periods when

$$p_t(x_t) = c'(e_t^{Th}) + \theta_t \leq \lambda \tag{5.19}$$

The shadow price on thermal capacity is positive only if the capacity is exhausted. If the social price is less than the water value then the water is saved to a period with a higher price. For a small enough share of hydro capacity of total capacity it may happen that hydro is used only in one period, the period with the highest price. Hydro may then become the typical peak-capacity power.

For periods where both hydro and thermal is used we have:

$$p_t(x_t) = \lambda = c'(e_t^{Th}) + \theta_t \tag{5.20}$$

In a situation with no period with a binding reservoir constraint and assuming that hydro will be used in every period the price will be constant for all periods.

Regarding the concepts base load and peak load it has been argued in Norway that investments should be made in thermal capacity to serve as peak load. On the other hand, a standard argument for a mixed hydro and thermal system is that hydro should be used as peak load because of its flexibility. Our analysis shows that without binding reservoir constraints, thermal capacity may be regarded as base load because it will be used at constant capacity (up to and including the maximal capacity) for all periods when hydro is also used, while the use of hydro will follow any shift of the demand over the periods. For periods that hydro is not in use the social price level must then be lower than the water value, implying that less thermal capacity will be used in such periods. In such a setting thermal capacity appears as base-load capacity and hydro as peak-load capacity. But such delimitation is rather crude when we operate with aggregate capacities. The concepts of peak and base load are more fruitfully applied at a disaggregated level showing individual generators.

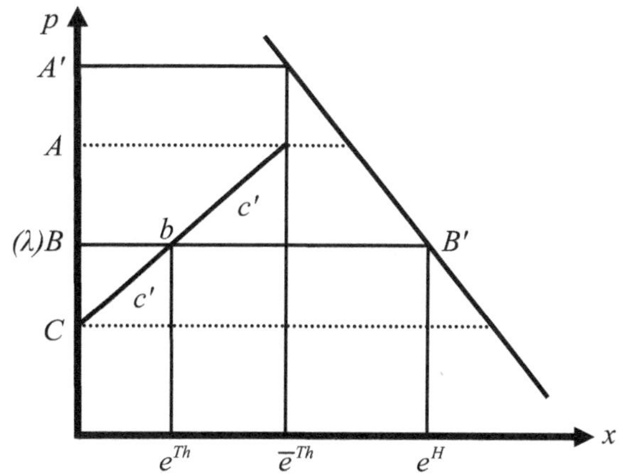

Figure 5.4. Hydro and thermal. Social optimum.

An illustration for one period of the use of the two technologies is shown in Figure 5.4. The marginal cost curve, $c'c'$, for thermal capacity starts at C and ends at the full capacity value, $\overline{e}^{Th}$. Assuming λ to be the water value, the optimal solution for the social price is at level B equal to the shadow price of water, and a thermal contribution of $Bb = e^{Th}$ and a hydro contribution of $bB' = e^{H}$. If we assume that the figure is representing just one of many periods it is meaningful to introduce two alternative water values by the dotted horizontal lines at levels C and A. For water values from levels A to A' the full capacity of thermal units will be utilised. For water values higher than at level A' only thermal capacity will be used. (Since we have the amount of water W to use up this situation cannot apply to all periods.) For water values lower than at level C no thermal capacity will be used. In a multi-period setting with identical demand functions and average availability of water being bB' the one period solution shown in the figure will be repeated each period.

For two periods we may expand the bathtub diagram to illustrate the allocation of the two types of power on the two periods. In Figure 5.5 the length of the bathtub AD is extended (analogous to the procedure in Figure 3.9 in Chapter 3) at each end with the thermal capacity. Dotted lines indicate the situation without thermal capacity. The demand curves after introduction of thermal capacity are anchored at the solid thermal "walls," i.e., horizontal shifts to the left, respectively right, for period 1 and 2. The marginal cost curve of thermal capacity is anchored at the broken hydro wall at $c'(0)$ to the left for period 1 and to the right for period 2. We assume the same cost curve for the two periods. The short vertical line at the end of

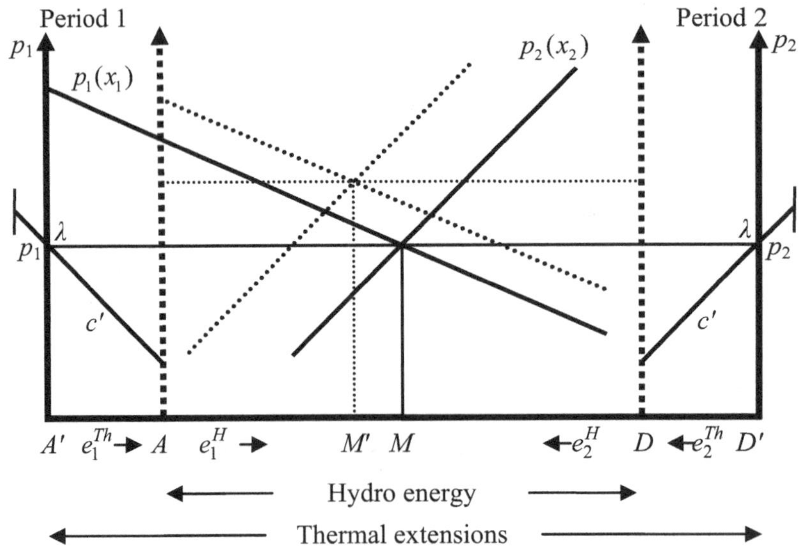

Figure 5.5. Energy bathtub with thermal-extended walls of the hydro bathtub. Solution with pure hydro shown by dotted lines.

the cost curves indicates the capacity limit. Using the result (5.20), we have that the thermal extension of the bathtub is equal at each end; with $A'A$ in period 1 and DD' in period 2 and $A'A = DD'$. The equilibrium allocation is at point M, resulting in an allocation of $A'A$ thermal and AM hydro in period 1, and MD hydro and DD' thermal in period 2 to the same social price, $p_1 = p_2$. In our example the allocation with thermal capacity results in less hydro used in period 2 when thermal capacity is also available, indicated by the allocation points M' and M for the situation without and with thermal capacity, respectively. The reason is that the demand in period 2 is more inelastic than for period 1. When introducing equal supply of thermal electricity in both periods in addition to hydro, the demand in period 1 increases more than the demand in period 2, because the demand in period 1 is more elastic than in period 2, leading to a *decreased* share of hydro in period 2. Changing the allocation on the two periods from M' to M we have that, since the shadow price for water and thereby the price becomes lower, the total electricity consumption increases in both periods.

Introducing a reservoir constraint

Introducing a reservoir constraint into problem (5.15) yields the following optimisation problem:

$$\begin{aligned}
&\max \sum_{t=1}^{T}[\int_{z=0}^{x_t} p_t(z)dz - c(e_t^{Th})] \\
&\text{subject to} \\
&x_t = e_t^H + e_t^{Th} \\
&R_t \le R_{t-1} + w_t - e_t^H \\
&R_t \le \overline{R} \\
&e_t^{Th} \le \overline{e}^{Th} \\
&x_t, e_t^H, e_t^{Th}, R_t \ge 0, \quad t = 1,..,T \\
&T, w_t, R_o, \overline{R}, \overline{e}^{Th} \text{ given}, R_T \text{ free}
\end{aligned} \tag{5.21}$$

The total hydro supply condition in (5.15) is replaced by the second and third condition in (5.21) showing the dynamics of water storage and the upper constraint on total storage. A constraint on hydro production capacity is not introduced here, but will be in the next section.

Inserting the energy balance the Lagrangian function is:

$$\begin{aligned}
L = &\sum_{t=1}^{T}[\int_{z=0}^{e_t^H + e_t^{Th}} p_t(z)dz - c(e_t^{Th})] \\
&-\sum_{t=1}^{T} \theta_t (e_t^{Th} - \overline{e}^{Th}) \\
&-\sum_{t=1}^{T} \lambda_t (R_t - R_{t-1} - w_t + e_t^H) \\
&-\sum_{t=1}^{T} \gamma_t (R_t - \overline{R})
\end{aligned} \tag{5.22}$$

The necessary first-order conditions are:

$$\frac{\partial L}{\partial e_t^H} = p_t(e_t^H + e_t^{Th}) - \lambda_t \le 0 \; (= 0 \text{ for } e_t^H > 0)$$

$$\frac{\partial L}{\partial e_t^{Th}} = p_t(e_t^H + e_t^{Th}) - c'(e_t^{Th}) - \theta_t \le 0 \; (= 0 \text{ for } e_t^{Th} > 0)$$

$$\frac{\partial L}{\partial R_t} = -\lambda_t + \lambda_{t+1} - \gamma_t \leq 0 \ (= 0 \text{ for } R_t > 0) \tag{5.23}$$
$$\lambda_t \geq 0 \ (= 0 \text{ for } R_t < R_{t-1} + w_t - e_t^H)$$
$$\gamma_t \geq 0 \ (= 0 \text{ for } R_t < \overline{R})$$
$$\theta_t \geq 0 \ (= 0 \text{ for } e_t^{Th} < \overline{e}^{Th}), \quad t = 1,..,T$$

Regarding combining hydro and thermal we will now have as a general rule that the water value is period specific in the first condition, implying that thermal capacity may vary between periods when both hydro and thermal capacities are used. From the second condition in (5.23) we have that the use of thermal capacity, when it is positive, but less than full capacity, is determined by equalisation of marginal costs and social price. The social price is equal to the water value for the period in question if hydro is used also. When the price varies due to threats of overflow and reservoir constraints being binding, as expanded upon in Chapter 3, then the use of thermal will vary with more capacity being used the higher the price, and thus a peak-load role follows also for thermal.

A possible situation is illustrated in Figure 5.6. The figure is built up in the same way as Figure 5.5. The total hydro capacity is AD with inflow

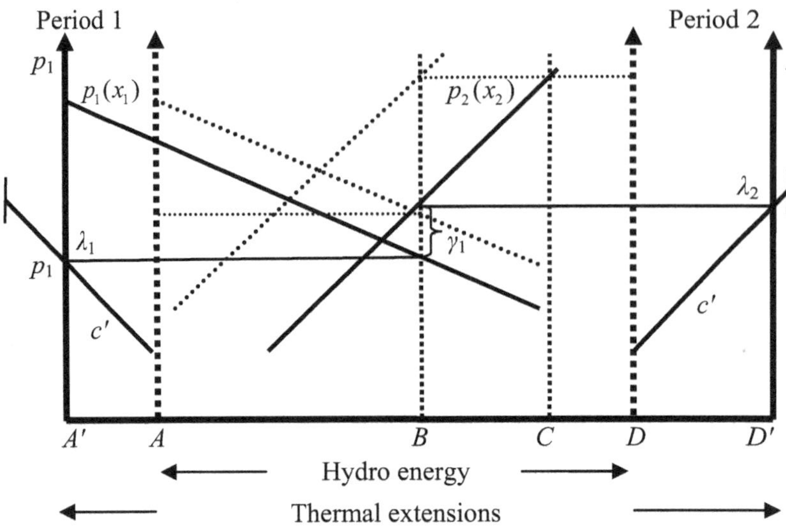

Figure 5.6. Thermal and hydro with reservoir constraint. Solution with pure hydro shown by dotted lines.

AC in period 1 and *CD* in period 2 and storage capacity is *BC*. The demand curves within the hydro bathtub without thermal capacity are indicated by thin dotted lines. The configuration of the demand curves is such that maximal water is transferred to period 2, and the price difference between the periods is considerable with hydro only, as indicated by the thin dotted horizontal hypothetical price lines to the hydro walls from the intersection points with the dotted demand curves and the vertical broken line erected in *B*. After introducing thermal capacity the maximal amount is still stored in period 1 for use in period 2. This means that the water allocation is unchanged between the periods. Since thermal capacity is not utilised to its maximum in any of the two periods the period water value should be set equal to the marginal thermal costs. This implies that less thermal capacity, *A'A*, is used in period 1 with the lowest water value, and more thermal capacity, *DD'*, is taken into use in the second period. We can say that the thermal capacity in period 1 is base load, and that the increase in output in period 2 is peak load. The price difference after introducing thermal capacity is considerably smaller (and, of course, both period prices are lower due to increased electricity supply). Other possible configurations of the optimal social solution in the multiperiod case may follow the discussion in Chapter 3.

Optimal mix of hydro and thermal plants

The previous section has been based on aggregated supply both from hydropower and thermal plants. But discussing the issue of peak load and base load is a little crude based on aggregate supply for hydro and thermal plants. Whether capacity serves peak or base load is a question characterising individual plants. We will investigate this topic by combining the multi-plant hydropower model of Chapter 4 with individual thermal plants. In addition to reservoir constraints, production constraints will also be specified for the hydropower plants, paralleling the treatment of thermal capacities.

The planning problem becomes:

$$\max \sum_{t=1}^{T} [\int_{z=0}^{x_t} p_t(z)dz - \sum_{i=1}^{M} c_i(e_{it}^{Th})]$$

subject to

$$x_t = \sum_{j=1}^{N} e_{jt}^{H} + \sum_{i=1}^{M} e_{it}^{Th}$$

$$\begin{aligned}
& R_{jt} \le R_{j,t-1} + w_{jt} - e_{jt}^{H} \\
& R_{jt} \le \overline{R}_{j} \\
& e_{jt}^{H} \le \overline{e}_{j}^{H} \\
& e_{it}^{Th} \le \overline{e}_{i}^{Th}, \\
& x_t, e_{it}^{H}, e_{jt}^{Th}, R_{jt} \ge 0, \\
& T, w_{jt}, R_{jo}, \overline{R}_j, \overline{e}_j^{H}, \overline{e}_i^{Th} \text{ given}, R_{jT} \text{ free}, \\
& i = 1,..,M,\ j = 1,..,N,\ t = 1,..,T
\end{aligned} \tag{5.24}$$

The first constraint is the energy balance adding up supply both from hydro and thermal plants. As mentioned in Chapters 3 and 4, the two last production constraints in (5.24) may also be interpreted as effect constraints. This is a more common practice for thermal plants. As noted earlier, the equivalence between production and effect constraints here is due to the basic assumption of using effect at a constant rate during the length of time period chosen.

Following our procedure of substituting for total consumption from the energy balance the Lagrangian for problem (5.24) becomes:

$$\begin{aligned}
L = & \sum_{t=1}^{T} \Big[\int_{z=0}^{\sum_{j=1}^{N} e_{jt}^{H} + \sum_{i=1}^{M} e_{it}^{Th}} p_t(z)dz - \sum_{i=1}^{M} c_i(e_{it}^{Th}) \Big] \\
& - \sum_{t=1}^{T} \sum_{j=1}^{N} \lambda_{jt}(R_{jt} - R_{j,t-1} - w_{jt} + e_{jt}^{H}) \\
& - \sum_{t=1}^{T} \sum_{j=1}^{N} \gamma_{jt}(R_{jt} - \overline{R}_j) \\
& - \sum_{t=1}^{T} \sum_{j=1}^{N} \rho_{jt}(e_{jt}^{H} - \overline{e}_j^{H}) \\
& - \sum_{t=1}^{T} \sum_{i=1}^{M} \theta_{it}(e_{it}^{Th} - \overline{e}_i^{Th})
\end{aligned} \tag{5.25}$$

The first-order necessary conditions are:

$$\frac{\partial L}{\partial e_{jt}^{H}} = p_t(\sum_{j=1}^{N} e_{jt}^{H} + \sum_{i=1}^{M} e_{it}^{Th}) - \lambda_{jt} - \rho_{jt} \le 0 (= 0 \text{ for } e_{jt}^{H} > 0)$$

$$\frac{\partial L}{\partial R_{jt}} = -\lambda_{jt} + \lambda_{j,t+1} - \gamma_{jt} \leq 0 (= 0 \text{ for } R_{jt} > 0)$$

$$\frac{\partial L}{\partial e_{it}^{Th}} = p_t (\sum_{j=1}^{N} e_{jt}^{H} + \sum_{i=1}^{M} e_{it}^{Th}) - c_i'(e_{it}^{Th}) - \theta_{it} \leq 0 \ (= 0 \text{ for } e_{it}^{Th} > 0)$$

$$\lambda_{jt} \geq 0 (= 0 \text{ for } R_{jt} < R_{j,t-1} + w_{jt} - e_{jt}^{H}) \tag{5.26}$$

$$\gamma_{jt} \geq 0 (= 0 \text{ for } R_{jt} < \bar{R}_j)$$

$$\rho_{jt} \geq 0 (= 0 \text{ for } e_{jt}^{H} < \bar{e}_j^{H})$$

$$\theta_{it} \geq 0 (= 0 \text{ for } e_{it}^{Th} < \bar{e}_i^{Th}), \quad i = 1,..,M, t = 1,..,T, j = 1,..,N$$

As mentioned before, the thermal cost functions must be well behaved, and start-up costs disregarded for standard sufficiency conditions of concavity to apply.

The qualitative discussion of management of single hydropower plants in Chapter 4 is valid also for combined hydro and thermal plants. The water values may become plant specific and social prices may change between periods, not only due to the dynamics of reservoir-related shadow prices, but also due to the interaction between production, including production from thermal plants, and aggregate demand. The management of thermal plants naturally does not involve explicitly the dynamic equation of the movement of reservoir-related shadow prices as shown in the second condition in (5.26). The running of thermal plants follows straightforwardly from the third condition in (5.26). Plants should not produce in periods where marginal cost at zero production exceeds the period price:

$$p_t (\sum_{j=1}^{N} e_{jt}^{H} + \sum_{i=1}^{M} e_{it}^{Th}) \leq c_i'(0), t = 1,..,T \tag{5.27}$$

Since the price is of crucial importance for whether a plant is operated or not, it would be tempting to associate peak load plants with high prices and base load plants with low prices. Let us first have a look at how prices covary with load. Figure 5.7 shows the development of hourly prices for Norway in 2005 when the hours are sorted from left to right according to the load curve Figure 1.3 in Chapter 1. It is a tendency for prices to be higher at peak load and get smaller toward the shoulder and base load part of the load curve, although the variation along the curve of the average level is large. The typical variation between hours is between 200 and 300 NOK per MWh (Euro 25-37), or a variation of 50%. This may be a surprisingly high variation of prices, but the hours may represent any time of

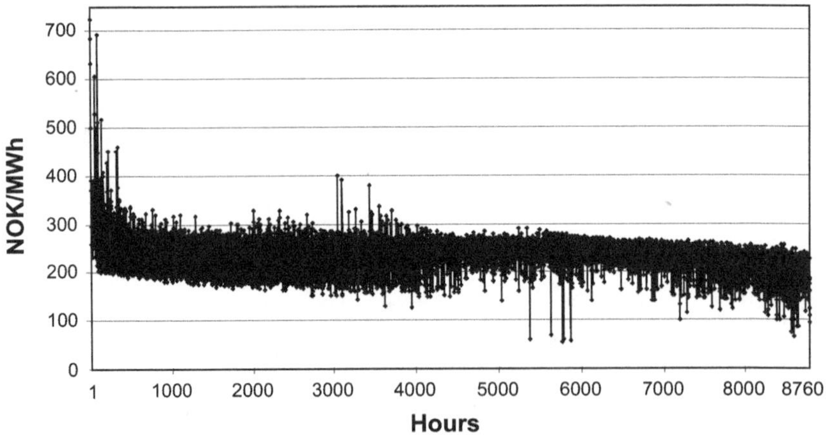

Figure 5.7. Hourly prices along the load-duration curve shown in Figure 1.3. Norway 2005.

the year and day. However, there is a tendency for price spikes to occur along the first half of the distribution of the highest load, and price troughs to occur along the latter half of the load-curve distribution. The highest price spike represents the third highest load, and of the 10 prices over NOK 500 (Euro 61), all but one price is in the morning hours between 0700 and 0900 hours, while the only exception is one price spike in the afternoon 1800 hours. All the high-price incidents occur in wintertime. However, the lowest price episodes are not toward the base-load end of the load curve coinciding with summer nights, holding for a majority of low prices, but a night time in November. It is probably due to the impact of autumn rain and unregulated hydropower.

Base-load plants are by a strict definition utilised in all periods. Considering thermal plants the third condition in (5.26) tells us that for base plants u being fully utilised and v being partly utilised in period t we have:

$$\begin{aligned} &c'_u(\bar{e}_u^{Th}) + \theta_{ut} = c'_v(e_{vt}^{Th}) = p_t(x_t) \\ &u \in U^t, v \in V^t, U^t \cup V^t = B^t, t = 1,..,T \end{aligned} \tag{5.28}$$

Here B^t is the set of base load plants in period t. If a plant, u, is utilised to its production capacity $\bar{e}_u^{Th}$ in period t, then the capacity shadow price typically becomes positive. Another plant, v, may not reach its capacity in period t and its shadow price therefore remains zero. Both types face the same social price for period t. The sets of thermal plants fulfilling one of the two situations in period t is U_t and V_t and the total set of base load plants in period t is B^t. Thermal plants fulfilling one of the two conditions in (5.28) (either being utilised below or at the capacity limit) for *all* time periods will be defined as base-load plants. A looser definition of base-load plants would be to focus on the share of time during a year that a plant delivers at the different segments of the load curve. A lower limit for inclusion in the category base load can be 50%. The water value may become so low that no thermal plants are operated all the time. Nuclear plants may be operated even in periods with water values lower than marginal costs because of high start-up and closing costs. Nuclear plants are therefore always run as base plants and are down only for scheduled (or unscheduled) service.

Peak-load thermal plants obey the same conditions (5.28) when they are operated. By definition base load plants are also run at peak-load periods. What distinguishes peak-load plants is that they are idle in other hours. Peak periods occur as only a certain fraction of total yearly hours. To classify a plant as a peak plant we have to delimit peak periods. We could go for a fraction of the periods with highest load, say, 20%, or we could use a fraction the period demand is over base load. If the set of peak-load hours is PT, then plants are defined as peak load when the following conditions are fulfilled:

$$\begin{aligned} &c_u'(\bar{e}_u^{Th}) + \theta_{ut} = c_v'(e_{vt}^{Th}) = p_t(x_t) \text{ for } t \in PT \\ &e_{ut}^{Th} = e_{vt}^{Th} = 0 \text{ for } t \notin PT,\ u, v = 1,.., M \end{aligned} \tag{5.29}$$

As for base load this definition could be weakened by allowing a peak plant to operate outside peak-load hours, but demand that the fraction of yearly output produced in peak periods should be, e.g., above 50%.

Shoulder load could be defined in a similar way as done in (5.29) for peak-load, but this category is usually not so much in focus, so this is left to the reader.

Figure 5.2 in the first section of the chapter illustrated the role of marginal costs in defining base and peak load plants. In the second section Figure 5.5 illustrated that more thermal capacity is used the higher the price. By operating with individual plants the location of them along an aggregated supply curve can be identified and the classification of base and peak load plants be made operational.

Since the key characteristic of hydropower plants is that they do not have variable costs, classifications into base and peak load may not be so interesting. The water values for hydro base loads plants do not have to be equal to the marginal costs of base load thermal plants, but the shadow prices on production capacities and water must adjust such that the respective sums add up to the social price for each period. Combining the first and the third condition in (5.26) assuming positive amounts of both hydro and thermal yields:

$$\lambda_{jt} + \rho_{jt} = c_i'(e_{it}^{Th}) + \theta_{it} = p_t, \, t \in B^t, \tag{5.30}$$

where the index j denotes hydro plants and index i thermal ones. While thermal marginal costs are *technically* given at the capacity limits the water values are determined in the process of finding the optimal solution to the planning problem. The management principle for hydropower plants, as expanded upon in Chapter 4, is to save as much water as possible to high price periods in order to maximise value creation. Hydro plants will not be used if the water value is higher than the current price. Then water will be accumulated for use in later higher-price periods taking production and reservoir constraints into due consideration. Plants with smaller storage capacities and/or more abundant inflows will tend to be producing in more periods than plants with large storage capacities. The reservoirs of these latter plants will typically be utilised during high-price periods. This is the role of plants in Norway with capacity to store several years with average inflows. But such plants may also produce in other periods due to performing the balancing act between inflows, reservoir and production capacity.

A dynamic thermal problem

As mentioned earlier, there are in practice adjustment costs associated with thermal plants. Structures and water must be warmed up and steam pressure built up before electricity production can take place if starting from a cold state. A start-up from a cold state may use other more costly energy-rich fuels than the ones used in a producing mode. A thermal station is in a spinning state when it is ready to produce, but still does not do so. This state also entails a cost, mainly in the form of burning of primary energy. We should also be aware of the fact that it may be technical problems with starting from spinning and produce just a marginal amount of electricity. Engineering information indicates that a plant has to be taken straight into a certain amount of the share of maximal output, maybe 1/4. (This may also be due to a concave marginal cost function.) When turning off a plant this operation in itself may entail some energy or labour cost, but most of the cost consists of loss of heat from warm structures and water. It will take some time before a plant is back to a cold state. Managing the plant taking these events into consideration implies that a dynamic problem must be solved. It does not seem to be so meaningful to pose these adjustment problems for the aggregate supply as captured by the cost function (5.8). It may be more relevant to face the problem at a plant level. But this also depends on the length of time period considered, hours within a day, days, weeks, etc.

We will develop a very simple example based on linear total cost functions as shown in Figure 5.2. Three plants are involved, representing peak, shoulder and base capacity. We will study these plants as if they operate in a "market," i.e., the period prices within a complex system also including hydropower generation come out as solutions to a social planning problem. Furthermore, the operation of these three plants does not influence the optimal price values. Three periods only are considered and the period price fluctuates between two values. The periods may be thought of as daytime and night-time. Figure 3.3 in Chapter 3 showed that the main variation in prices is between daytime and night-time levels. The problem is set up in such a way that it is a question only about whether to close down the peak load capacity in period 2 (night time) or not. It is clear that base and shoulder capacity should be run at full capacity for all the three periods. The start-up costs of the period before the first period are neglected. The situation is portrayed in Figure 5.8. The step-curve $b_1b_2b_3$ is the supply curve and the capacities of each technology are indicated on the horizontal axis. The price fluctuates from the value of the upper price line in period 1 (daytime) to the value of the lower price line in period 2 (night time) and back again in period 3. The adopted cost functions (5.3) are:

$$c_{it} = c_i(e_{it}^{Th}) = a_i + b_i e_{it}^{Th},\ i,t = 1,..,3\,, \tag{5.31}$$

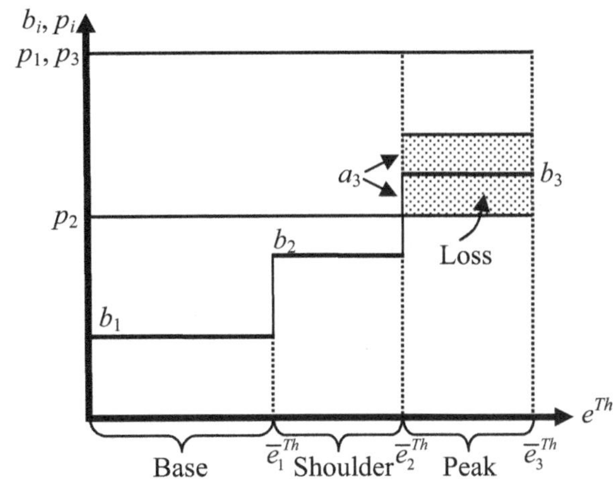

Figure 5.8. Start-up costs.

where a_i is the start-up cost of plant i incurred if the plant is switched off in a period and wants to start up again in the next (or a later) period. It is assumed that stopping at the end of period 1 and starting up at the beginning of period 3 is technically feasible. The time variations of the heat loss and spinning-state costs are neglected. The peak load plant 3 incurs an operating loss in period 2 if it is running equal to the lower marked area in the figure. The start-up cost if the plant is shut down is the area a_3 marked as the sum of the two hatched areas in the figure. The partial management problem for the social planner is to inspect the best action of either temporarily shutting down plant 3 in period 2 and then start it up for period 3, or to let the plant run in period 2 and incur a loss, but to avoid facing start-up costs in period 3. Notice that the close-down decision and the start-up decision must be taken simultaneously. The optimal decision depends on the size of the start-up costs and the profit in period 3 under the alternatives. If the start-up cost is greater than the loss incurred by having the plant running in period 2, then it cannot be optimal to shut it down in period 2 and start it up again in period 3. But the condition to keep it running is that the profit made in period 3 is greater than the loss incurred in period 2. The conditions for choosing to run plant 3 with a loss in period 2, and then continue to run it in period 3 are:

$$\begin{aligned} &a_3 > |p_2 - b_3| \bar{e}_3^{Th} , \\ &p_3 - b_3 - (b_3 - p_2) = p_3 + p_2 - 2b_3 > 0 \Rightarrow \frac{p_2 + p_3}{2} > b_3 \end{aligned} \tag{5.32}$$

The expression on the right-hand side of the first condition is the absolute value of the operating loss and is the lower marked rectangle in Figure 5.8. The condition is clearly fulfilled in the figure. The second condition requires that it is profitable to run the unit in period 3, i.e., the operating surplus in period 3 must be able to absorb the loss in period 2. This is fulfilled if the average price of the two periods is higher than the marginal cost, which is the case in the figure.

If the start-up cost is less than the operating loss in period 2, then the plant should be closed down in period 2 (i.e., not be running) and started up again in period 3, provided the operating surplus can also absorb the start-up costs:

$$a_3 < |p_2 - b_3| \bar{e}_3^{Th} , \ (p_3 - b_3)\bar{e}_3^{Th} - a_3 > 0 \tag{5.33}$$

If the number of periods is increased, if a plant is stopped, then it may be reactivated the first period the price is higher than marginal costs, although the whole start-up cost does not need to be recouped in this period if there are enough successive periods with positive quasi-rent to recover the start-up cost.

It may well be that so many thermal plants are involved in adjustments described above that the optimal prices may be influenced in the planning problem. In the case the planner finds that plants should produce although they are incurring losses, the equilibrium price will be influenced downwards, and in the case of closing down temporarily there may be an upward pressure on prices in succeeding periods until the (generalised) second condition in (5.33) is met.

If the time period definition is not too aggregated spinning costs are typically lower than start-up costs. However, if the optimal decision is to close down the peak plant without considering spinning, then spinning will not be an alternative if the plant can be started up immediately in the next period. If this assumption is changed the situation may become different. It may be realistic that it takes some time to plan and prepare for activating a plant from a cold state. If this fact should be modelled depends on the length of the time period in question. Let us assume that it takes two periods to start up again from a cold state, but that starting to produce from spinning is immediate, per definition. Then if we look at more periods than three as in the example above, and furthermore assume that prices after the second period, where there is an operational loss, allows operational surplus, then spinning in the second period may become optimal. Closing down the plant implies that the positive quasi-rent in the third period is lost, since it takes two periods to reopen the plant. The condition for spinning is:

$$\begin{aligned} &(p_3 - b_3)\bar{e}_3^{Th} + (p_4 - b_3)\bar{e}_3^{Th} - s_3 > (p_4 - b_3)\bar{e}_3^{Th} - a_3 \Rightarrow \\ &a_3 > s_3 - (p_3 - b_3)\bar{e}_3^{Th} \end{aligned} \tag{5.34}$$

Here s_3 is the spinning costs of plant 3. If periods after period 4 are all surplus periods, then they cancel out in the calculation above. If spinning costs are lower than start-up costs, then in this situation spinning will always be preferred. If we change the assumption of positive operating surplus in period 3, but maintain positive surplus in period 4 and later periods, then the condition for spinning being optimal is:

$$\begin{aligned} &(p_4 - b_3)\bar{e}_3^{Th} - 2s_3 > (p_4 - b_3)\bar{e}_3^{Th} - a_3 \Rightarrow \\ &a_3 > 2s_3 \end{aligned} \tag{5.35}$$

It can now happen that is more profitable to close down a plant rather than keeping it spinning, depending on whether the spinning cost is more than half the start-up cost. It is straightforward to introduce other assumptions about the length of the start-up lag and profile of the operating surplus.

Chapter 6. Trade

We can think about physical trade in electricity at two levels of aggregation: between countries and between regions within a country. Starting with the latter level the models with individual plants in Chapter 4 are examples. However, the trade flows were not specified there. This is not so conveniently done operating without an explicit transmission system. A transmission system is introduced in Chapter 7. In order to study trade between two countries a single interconnector only will be assumed. The aggregate treatment of the hydropower sector can then be maintained and the analysis can be conducted without specifying transmission and still bring out some main points.

As pointed out in Chapter 1, isolating a country from trade in electricity creates a country-specific price that may influence the structure of industry and, e.g., choice of space-heating technology. This has been the case for Norway developing a huge metal smelting industry after World War II, also in an international context, and basing a significant share of space heating on direct use of electricity. It is therefore of interest to study what happens with the price formation at home when borders are opened up for trade in electricity. There is a common international market, Nord Pool, between the Nordic countries since 1996, and international trade now takes place between many European countries on a bilateral basis, e.g., France – England, France – Italy (Italy imports about 20% of its electricity), etc. The energy policy of the European Union is encouraging a gradual expansion of cross-border trading and integration of electricity markets (Jamasb and Pollit, 2005).

Unconstrained trade

Introducing trade means that we introduce a second good, money, into our model country in addition to electricity. We will simplify by just adding (subtracting) the export (import) in money to (from) the area under the demand curve for electricity, implying that in the background we assume utility functions separable in electricity and money (an aggregate for all other goods). We will start by assuming only hydropower in the home country. The objective function will then be the sum over the periods of consumer and producer surplus, which in our case for electricity will be the gross area under the demand curve since we have assumed zero production cost (only water value counts), and for money there is just the amount: positive for exports and negative for import. Our model is partial, so we have no constraint on the balance of trade in electricity. It may well be an optimal solution to import for more than we earn in export provided the increase in the area under the demand curves more than compensates for an eventual deficit on the electricity trade. We are not concerned about balance of trade for the total economy that may be implicitly assumed in the background.

The country energy balance now involves export and import:

$$x_t = e_t^H - e_t^{XI}, t = 1,..,T \tag{6.1}$$

The variable e_t^{XI} is net export or import and is positive if we have export and negative if we have import. We assume that in one period we can only have either export or import, or both can be zero. There is no restriction to have balance of trade in electricity, as mentioned above.

The social planning problem studied first is how to manage hydropower resources when a country has access to unlimited trade in electricity to given prices, and reservoir limits and other constraints on transmission and production are disregarded:

$$\begin{aligned}
&\max \sum_{t=1}^{T}[\int_{z=0}^{x_t} p_t(z)\,dz + p_t^{XI} e_t^{XI}] \\
&\text{subject to} \\
&x_t = e_t^H - e_t^{XI} \\
&\sum_{t=1}^{T} e_t^H \le W \\
&x_t, e_t^H \ge 0,\ \ e_t^{XI} \text{ unrestricted} \\
&T, W, p_t^{XI} \text{ given, } t = 1,..,T
\end{aligned} \tag{6.2}$$

The transmission system is still not shown explicitly. It is assumed that there is enough transmission capacity for the trade volumes in question. We could assume a certain fixed cost per unit transmitted, but this will not change our analysis, so we will assume that the import price is equal to the export price. These prices are given and not influenced by actions of our country. In the last section models are developed in which export/import prices are endogenously determined.

Substituting for total consumption from the energy balance in the objective function, the Lagrangian becomes:

$$L = \sum_{t=1}^{T}[\int_{z=0}^{e_t^H - e_t^{XI}} p_t(z)dz + p_t^{XI} e_t^{XI}] \\ -\lambda(\sum_{t=1}^{T} e_t^H - W) \tag{6.3}$$

The necessary first-order conditions are:

$$\frac{\partial L}{\partial e_t^H} = p_t(e_t^H - e_t^{XI}) - \lambda \le 0 \; (= 0 \text{ for } e_t^H > 0) \\ \frac{\partial L}{\partial e_t^{XI}} = -p_t(e_t^H - e_t^{XI}) + p_t^{XI} = 0 \\ \lambda \ge 0 \; (= 0 \text{ for } \sum_{t=1}^{T} e_t^H < W), \quad t = 1,..,T \tag{6.4}$$

It is quite reasonable to assume that electricity is provided to our country in every period; $x_t > 0$ for all $t = 1,\ldots,T$. This means that in export periods hydro is also used for home consumption and the first condition in (6.4) holds with equality. The second condition holds as an equality since there is no restriction on the sign of e_t^{XI}. The condition states that the export/import prices will be completely adopted as domestic prices. With no restriction on transmission or storage of water an important conclusion for prices is immediately that the foreign price regime will be adopted as the home country price regime.

Now, since the shadow price on water is without period subscript we can have only *one* export period if we make the assumption that all the export/import prices are different. The shadow price on water is, via the second condition in (6.4), set equal to this maximum price:

$$\lambda = \max_{t=1,..,T}\{p_t^{XI}\} \tag{6.5}$$

But notice that we do not necessarily use hydropower in all periods. If the price in the home market is less than the shadow price λ on water, no water shall be used for hydropower production in that period; we just import. Without any constraint on the possibility to store water the model is thus rather extreme because we will only export in *one* period, the period with the highest export price, and import in *all* other remaining $T-1$ periods to the going import price.

The total export will be:

$$e_{t*}^{XI} = W - x_{t*} \text{ , } x_{t*} = p_{t*}^{-1}(p_{t*}^{XI}) \tag{6.6}$$

where t^* is the period with the maximal export price defined in (6.5).

An illustration is provided for two periods employing the energy bathtub presented in Figure 6.1. The social management problem is how to use the given water within the two periods when there are unlimited import and export possibilities to given prices. The autarky solution is indicated by the prices, $p_1^{AU} = p_2^{AU}$ as shown by the horizontal thin dotted lines in accordance with the results of model (2.4) in Chapter 2. The allocation point on the electricity bathtub floor is A^{AU}. The period 2 trading price is set higher than the period 1 price, so according to our general results above no water is used in period 1, but all in period 2. In period 1 the demand for electricity is satisfied by import determined by the intersection of the horizontal trading price line p_1^{XI} and the demand curve for period 1, bringing us to point *C* on the electricity axis. The total import is *AC*. In period 2 all the water is processed and allocated between export and home consumption according to the intersection between the horizontal trade price line p_2^{XI} and the demand curve for period 2, bringing us to point *B* on the electricity bathtub floor. Export is *AB* and home consumption *BD*. The water value becomes equal to the trading price in period 2. Compared with the autarky

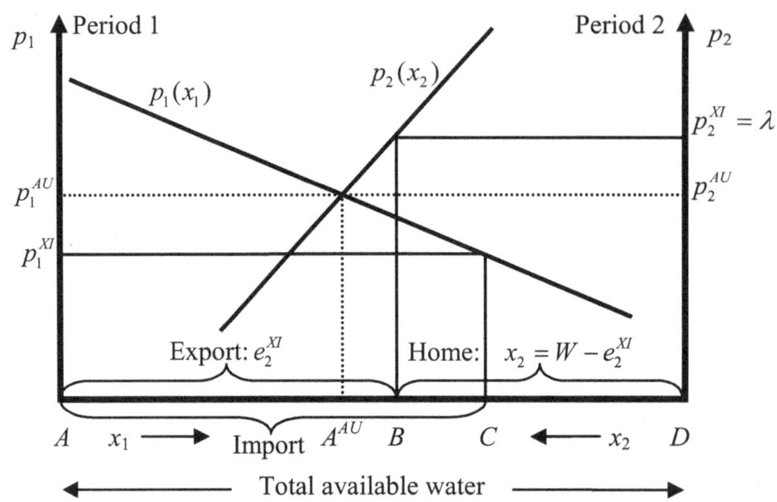

Figure 6.1. Unlimited trade. Autarky indicated by dotted lines.

solution, more electricity is consumed in period 1 and less in period 2. By comparing areas we should be able to see that the objective function has a higher value after trade. (Remember that the social manager can always choose to disregard trade.) Resources that are used in the economy for import and are obtained by exports are all measured in the same unit of money. But there may be some distributional issues hidden behind the aggregate results. We cannot know if the consumers facing higher prices and lower electricity consumption in period 2 are the same that benefits from low price and high consumption in period 1. The distribution of the export income, import expenditure, and financing of an eventual deficit of the electricity trade will also enter the picture.

Trade may be only of practical interest together with constraints on the volume of trade and/or the possibility to store water. Both modifications will be introduced in the next sections.

Reservoir constraint

The feature of only one export period and all other periods being import periods may seem too extreme. By introducing a reservoir constraint the unconstrained trade may get a more normal pattern. Replacing the total water constraint in (6.2) with the reservoir accumulation equation and the reservoir constraint the planning problem becomes:

$$
\begin{aligned}
&\max \sum_{t=1}^{T} [\int_{z=0}^{x_t} p_t(z)\,dz + p_t^{XI} e_t^{XI}] \\
&\text{subject to} \\
&x_t = e_t^H - e_t^{XI} \\
&R_t \le R_{t-1} + w_t - e_t^H \\
&R_t \le \overline{R} \\
&x_t, e_t^H, R_t \ge 0,\ e_t^{XI} \text{ unrestricted} \\
&T, w_t, p_t^{XI}, R_o, \overline{R} \text{ given},\ \ t = 1,..,T
\end{aligned}
\tag{6.7}
$$

The corresponding Lagrangian, substituting for total consumption from the energy balance in the objective function, is:

$$
L = \sum_{t=1}^{T} (\int_{z=0}^{e_t^H - e_t^{XI}} p_t(z)\,dz + p_t^{XI} e_t^{XI})
$$

$$-\sum_{t=1}^{T}\lambda_t(R_t - R_{t-1} - w_t + e_t^H)$$
$$-\sum_{t=1}^{T}\gamma_t(R_t - \overline{R}) \tag{6.8}$$

The first-order conditions are:

$$\begin{aligned}
&\frac{\partial L}{\partial e_t^H} = p_t(e_t^H - e_t^{XI}) - \lambda_t \le 0 \quad (=0 \text{ for } e_t^H > 0) \\
&\frac{\partial L}{\partial e_t^{XI}} = -p_t(e_t^H - e_t^{XI}) + p_t^{XI} = 0 \\
&\frac{\partial L}{\partial R_t} = -\lambda_t + \lambda_{t+1} - \gamma_t \le 0 \; (=0 \text{ for } R_t > 0) \\
&\lambda_t \ge 0 \; (=0 \text{ for } R_t < R_{t-1} + w_t - e_t^H) \\
&\gamma_t \ge 0 \; (=0 \text{ for } R_t < \overline{R}) \;, t = 1,..,T
\end{aligned} \tag{6.9}$$

As in the case of a total water constraint, the home country will adopt the trade prices, seen from the second condition. The new feature of introducing a reservoir constraint is that limits on using water for export will reduce transfers from import to export periods. The water values may become different. If the reservoir condition allows it there may be import periods without use of water.

The situation is illustrated in the two-period case in Figure 6.2. The available water in period 1 is *AC* and the broken vertical lines from *B* and *C* indicate the reservoir capacity *BC*. In the autarky situation indicated by thin dotted lines the prices become equal for the two periods ($p_1^{AU} = p_2^{AU}$) and the reservoir capacity is not fully utilised. With the chosen trade prices the full reservoir capacity is now used to transfer water to the highest price period 2. In that period export takes place. Domestic consumption is competing with exports resulting in *B'D* being consumed at home and *BB'* exported. Period 1 with the lowest trade price becomes the import period, and the intersection of the price line and the demand curve for period 1 determines the total consumption, *AC'*. But not all is imported, only *BC'*. There is an amount of water *AB* that is locked in due to the limited transferability, and has to be consumed at home. The water values become different with the lowest value in period 1 with forced consumption of hydropower. The difference between the water values is shown in the figure and is the shadow price on the reservoir constraint.

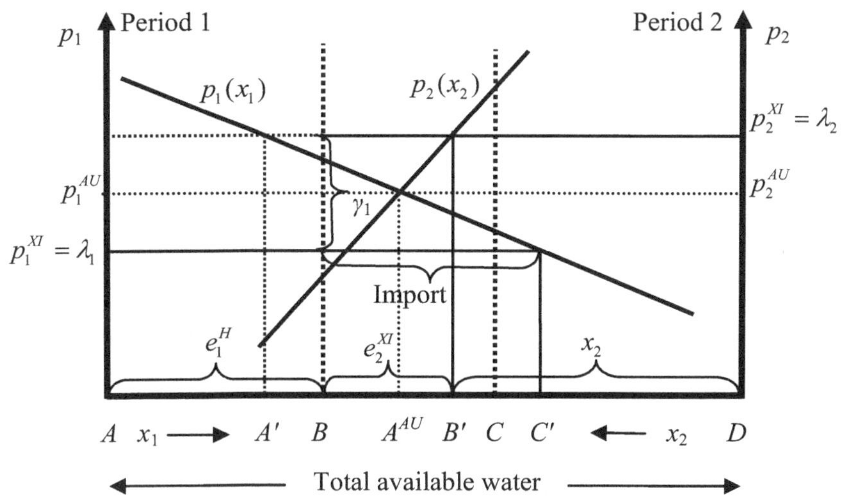

Figure 6.2. Unlimited trade with reservoir constraint. Autarky indicated by dotted lines.

There may now be several export periods in the general multi-period case. In the case of the trade prices being equal in Figure 6.2 and equal to the highest price, period 1 will become the export period and nothing will be imported. The amount of export in period 1 is determined by the intersection of the broken continuation of the price line to the left and the demand curve for period 1. The amount AA' will be consumed at home and $A'B'$ exported. The amount $B'C$ will be transferred to period 2, where $B'D$ will be consumed at home and nothing exported. The export price will be the home price for both periods and is equal to the common water value.

Constraints on trade

We now introduce an upper constraint on the volume of export/import. This constraint may take care of the capacities of the interconnection to the external market. Returning to the total water constraint, the social planning problem becomes:

$$\max \sum_{t=1}^{T} [\int_{z=0}^{x_t} p_t(z)dz + p_t^{XI} e_t^{XI}]$$

subject to

$$
\begin{aligned}
& x_t = e_t^H - e_t^{XI} \\
& \sum_{t=1}^{T} e_t^H \le W, \\
& -\overline{e}^{XI} \le e_t^{XI} \le \overline{e}^{XI} \\
& x_t, e_t^H \ge 0, e_t^{XI} \text{ unrestricted in sign} \\
& T, W, p_t^{XI}, \overline{e}^{XI} \text{ given } , t = 1,..,T
\end{aligned}
\tag{6.10}
$$

The constraint on trade can be split up into export and import. Since import by convention is negative a minus sign is put in front of the trade limit $\overline{e}^{XI}$ when import is constrained.

The corresponding Lagrangian substituting for total consumption from the energy balance in the objective function is:

$$
\begin{aligned}
L = & \sum_{t=1}^{T} \Big(\int_{z=0}^{e_t^H - e_t^{XI}} p_t(z)\,dz + p_t^{XI} e_t^{XI} \Big) \\
& - \lambda \Big(\sum_{t=1}^{T} e_t^H - W \Big) \\
& - \sum_{t=1}^{T} \alpha_t (e_t^{XI} - \overline{e}^{XI}) \\
& - \sum_{t=1}^{T} \beta_t (-e_t^{XI} - \overline{e}^{XI})
\end{aligned}
\tag{6.11}
$$

where α_t is the shadow price on the export constraint, $e_t^{XI} \ge 0$, and β_t is the shadow price on the input constraint, $e_t^{XI} \le 0$.

The first-order conditions are:

$$
\begin{aligned}
& \frac{\partial L}{\partial e_t^H} = p_t(e_t^H - e_t^{XI}) - \lambda \le 0 \quad (= 0 \text{ for } e_t^H > 0) \\
& \frac{\partial L}{\partial e_t^{XI}} = -p_t(e_t^H - e_t^{XI}) + p_t^{XI} - \alpha_t + \beta_t = 0 \\
& \lambda \ge 0 \ (= 0 \text{ for} \sum_{t=1}^{T} e_t^H < W) \\
& \alpha_t \ge 0 \ (= 0 \text{ for } e_t^{XI} < \overline{e}^{XI}) \ (e_t^{XI} > 0) \\
& \beta_t \ge 0 \ (= 0 \text{ for } -e_t^{XI} < \overline{e}^{XI}) \ (e_t^{XI} < 0), t = 1,..,T
\end{aligned}
\tag{6.12}
$$

We assume as before that consumption of electricity is positive in all periods; $x_t > 0$ for all $t = 1,..,T$. If there is export then $e_t^H > 0$ and the first equation in (6.12) holds with equality. The second equation holds with equality since export/import can be both positive and negative. Only one of the shadow prices on maximal trade can be positive in the same period (both can be zero). We have that if both shadow prices are zero (import/export constraints are not binding), then the home price is equal to the export/import price. But as opposed to the case without a restriction on the trade volume there is now no automatic adoption of the export/import prices domestically.

Let us again assume that all the export/import prices are different. Then there can only be *one* export period for which the upper trade constraint is not binding. The reason is that the shadow value on water has no time subscripts, and since the export prices are different we will have a contradiction with more than one such export period. Let us call the period for a *marginal export period.* There may be several export periods when the export constraint is binding. If the constraint on export is binding, then we may have that the export price is higher than the home price because we have in general from (6.12):

$$p_t(x_t) = \lambda = p_t^{XI} - \alpha_t \; (e_t^{XI} > 0) \tag{6.13}$$

A positive shadow price on the export constraint implies a lower home price than the export price.

For import periods we see from the first condition in (6.12) that we may have $e_t^H = 0$ if the home price is less than the shadow price on water for zero hydro production. We have in general for import periods

$$p_t(x_t) = p_t^{XI} + \beta_t \; (e_t^{XI} < 0) \tag{6.14}$$

If we are at the upper constraint for import with a positive shadow price β_t then the home price will typically be higher than the import price. Hydro can be used in import periods only if the transmission constraint is binding and the shadow price on the constraint is positive. The reason is that use of hydro with import below the trade constraint implies equality in the first condition in (6.12), and since import prices are different we will again have a contradiction. If we have an import period without a binding import constraint, the first condition in (6.12) tells us that the shadow price on water is higher than the home price, since there is zero hydropower production, and we then have from (6.14) that the home price adapts to the varying import price since the shadow value on the import constraint is zero. The number of import periods is determined residually when the number of export periods is determined.

A feasible optimal solution is illustrated in the two-period case in Figure 6.3. The hydro bathtub is extended on the left-hand side from the old hydro wall, indicated with a vertical dotted line, with the import in period 1, resulting in the new solid wall as the left-hand axis. By design this is the full capacity import. The shadow price β_1 on the import constraint in period 1 is indicated as the difference between the import price in period 1 and the home price. The difference between the sales value of the import and the import cost may be called the *congestion rent* and is equal to the product of the import capacity and the shadow price on the import constraint $\beta_1 \bar{e}^{XI}$ indicated by the marked rectangle.

In addition to import some hydro will also be used in period 1. The common shadow price on water is set equal to the highest trade price, occurring in period 2. In this period we have that the export is less than the transmission capacity. The home price is therefore equal to the export price in this period. Period 1 home price will also be the same because the opportunity value of water in period 1 is to export in period 2 since there is capacity to do so.

Disregarding limit on trade, we get the same result as in Figure 6.2 by using the dotted demand curve for period 1, anchored on the dotted left-hand hydro wall, and finding the intersection with the price line for the import price in period 1 (outside the right-hand bathtub wall). The import

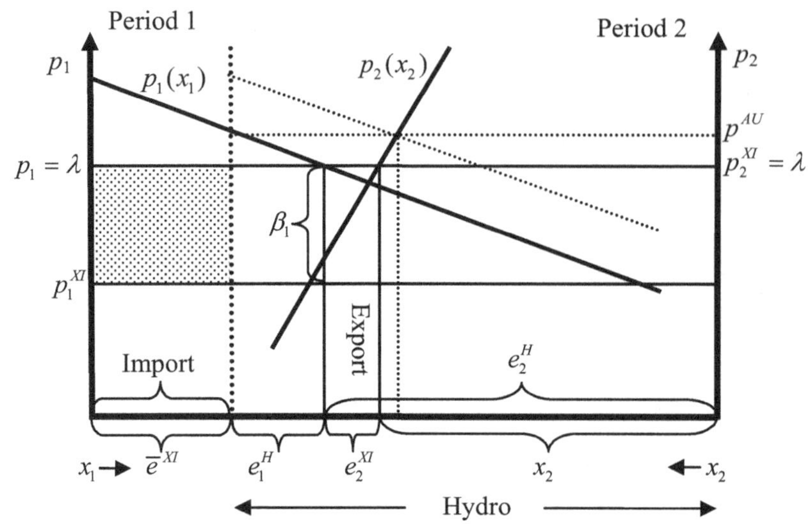

Figure 6.3. Limit on transmission capacity for trade. Autarky indicated by dotted lines.

would then be more than the total available hydro, and no hydro would now be used in period 1. In period 2 the same quantity of hydro would be consumed at home ($x_2 = e_2^H - e_2^{XI}$), but the export is extended with the amount of hydro used in the import period 1 with transmission constraint binding ($e_2^{XI}{}_{|\text{no restr.}} = e_1^H + e_2^{XI}$). Remember that we have no requirement of trade balance in electricity.

Compared with the autarky solution we have that both trade prices are lower than the common autarky price indicated by the horizontal dotted line p^{AU} in the figure. As to the allocation of water, the dotted vertical allocation line indicates that in autarky slightly more than home consumption of water in the import period plus the export in period 2 will now be consumed in period 1, resulting in somewhat less consumption of electricity than with trade. The consumption in period 2 under trade is just a little less than under autarky. What we see when restricted trade is introduced is that the difference in trade prices is utilised in order to shift water previously used in period 1 to export in period 2, or said in another way, utilising the cheapest trade-price period to import and save water, and then export in the high-price period. In the process home prices fall, indicating higher home consumption in both periods. The consumption is especially higher in period 1, not only because it is the import period, but also because the demand is more price elastic in period 1.

Returning to the general multiperiod case, many different trade patterns may emerge. Let us simplify by sorting the export/import prices in descending order and assuming that they are all different so we have a unique ranking. With no constraints on the volume of trade we found that export will take place in only one period, the maximal price period, and there will be import on all the other periods. We will also now have export in the highest export price period, but if it is assumed that the transmission constraint will become binding in this period with the highest price, then there will be export in at least one more period, depending on the relationship between the total amount of water, water used in export and import periods, and the constraint on trade, $\overline{e}^{XI}$. In the single export period when the constraint will typically not be binding (the case of all export periods hitting the constraint is quite arbitrary) the price for this period is the lowest among the set of prices for export periods. This price, $p_{t^*}^{XI\min}$ in period t^*, will then determine the shadow price λ on water. This the marginal export period.

As mentioned above (6.14), if water is used in an import period, it means that the import constraint is binding, and that at the import price in question, there is a positive residual home demand that can be satisfied only by using water. Since the alternative use of water is to increase export in the marginal export period, this implies that the home price in an import period with the transmission constraint binding must be equal to the water value and equal to the export price in the marginal export period. Conditional on knowing $p_{t*}^{XI\min}$ the set of periods with both imports and use of hydro at home can be defined:

$$T^{H+imp} = \{t : p_t\big|_{e_t^{XI}<0, e_t^H>0} = (p_t^{XI} + \beta_t)\big|_{e_t^{XI}<0, e_t^H>0} = p_{t*}^{XI\min}\big|_{e_{t*}^{XI}>0} = \lambda\} \tag{6.15}$$

The optimal shadow price on water λ must satisfy the condition that the total available water, W, is just used up on home consumption and exports [see Equation (6.18) below].

The set of periods with imports only and no use of water at home is defined by:

$$T^{imp} = \{t : p_t\big|_{e_t^{XI}<0, e_t^H=0} = p_t^{XI}\big|_{e_t^{XI}<0, e_t^H=0}\} \tag{6.16}$$

Conditional on knowing $p_{t*}^{XI\min}$ the set of periods with exports can be defined by:

$$T^{ex} = \{t : p_t^{XI} \geq p_{t*}^{XI\min}\big|_{e_{t*}^{XI}>0}\} \tag{6.17}$$

We have that the set of all T periods is the sum $T^{imp} \cup T^{H+imp} \cup T^{ex}$. The number of export periods, t^{ex} (an integer number), is found by looking at the balance of water supply and demand consisting of export and home consumption over all periods:

$$\begin{aligned} W &= (t^{ex} - 1)\bar{e}^{XI} + \sum_{t \in T^{ex}} x_t + e_{t*}^{XI} + \sum_{t \in T^{H+imp}} e_t^H \Rightarrow \\ t^{ex} &= \frac{W - \sum_{t \in T^{ex}} x_t - e_{t*}^{XI} - \sum_{t \in T^{H+imp}} e_t^H}{\bar{e}^{XI}} + 1 \end{aligned} \tag{6.18}$$

We have that $t*$ is the single export period when export is not hitting the upper constraint, T^{H+imp} is the set of import periods when hydro is also used, and T^{ex} is the set of export periods. The t^{ex} numbers of highest prices will belong to the export periods, and the rest of the prices will belong to import periods. In the $t^{ex} - 1$ number of periods with the highest prices the transmission constraint will be binding and typically the shadow price α_t is positive, driving a wedge between the lower home price and the export prices. As remarked above all the home prices for export periods and periods with both hydro and import [Equation (6.15)] are equal, so the shadow prices on the transmission constraint will all be different. In the period with the price ranked as number t^{ex} the export constraint is not binding and then the home price and the export price are equal and equal to the shadow price λ on water. In the $(T - t^{ex})$ periods with the prices lower than $p_{t^*}^{XI\min}$ we will have import and no use of hydro when the transmission constraint is not binding and use of hydro in addition when the transmission constraint is binding with positive shadow price.

Reservoir constraints

The most realistic case is to have a restriction both on interconnector capacity and on the reservoir in the home country. The resulting trade pattern would then conform better with what we observe. Introducing a reservoir in (6.10) the social optimisation problem becomes:

$$
\begin{aligned}
&\max \sum_{t=1}^{T}\left[\int_{z=0}^{x_t} p_t(z)\,dz + p_t^{XI} e_t^{XI} \right] \\
&\text{subject to} \\
&x_t = e_t^H - e_t^{XI} \\
&R_t \le R_{t-1} + w_t - e_t^H \\
&R_t \le \overline{R} \\
&-\overline{e}^{XI} \le e_t^{XI} \le \overline{e}^{XI} \\
&x_t, e_t^H, R_t \ge 0\,,\ e_t^{XI} \text{ unconstrained in sign} \\
&T, w_t, R_o, \overline{R}, p_t^{XI}, \overline{e}^{XI} \text{ given, } R_T \text{ free, } t = 1,..,T
\end{aligned}
\tag{6.19}
$$

The two restrictions after the energy balance substitute for the single total water constraint in problem (6.10).

Substituting for consumption from the energy balance into the objective function the Lagrangian function is:

$$
\begin{aligned}
L = &\sum_{t=1}^{T}\left(\int_{z=0}^{e_t^H - e_t^{XI}} p_t(z)\,dz + p_t^{XI} e_t^{XI} \right) \\
&-\sum_{t=1}^{T} \lambda_t (R_t - R_{t-1} - w_t + e_t^H) \\
&-\sum_{t=1}^{T} \gamma_t (R_t - \overline{R}) \\
&-\sum_{t=1}^{T} \alpha_t (e_t^{XI} - \overline{e}^{XI})
\end{aligned}
\tag{6.20}
$$

$$-\sum_{t=1}^{T}\beta_t(-e_t^{XI}-\overline{e}^{XI})$$

The necessary first-order conditions are:

$$\begin{aligned}
&\frac{\partial L}{\partial e_t^H}=p_t(e_t^H-e_t^{XI})-\lambda_t\le 0 \quad (=0 \text{ for } e_t^H>0)\\
&\frac{\partial L}{\partial e_t^{XI}}=-p_t(e_t^H-e_t^{XI})+p_t^{XI}-\alpha_t+\beta_t=0\\
&\frac{\partial L}{\partial R_t}=-\lambda_t+\lambda_{t+1}-\gamma_t\le 0 \ (=0 \text{ for } R_t>0)\\
&\lambda_t\ge 0 \ (=0 \text{ for } R_t<R_{t-1}+w_t-e_t^H)\\
&\gamma_t\ge 0 \ (=0 \text{ for } R_t<\overline{R})\\
&\alpha_t\ge 0 \ (=0 \text{ for } e_t^{XI}<\overline{e}^{XI}) \ (e_t^{XI}>0)\\
&\beta_t\ge 0 \ (=0 \text{ for } -e_t^{XI}<\overline{e}^{XI}) \ (e_t^{XI}<0)\ ,\ t=1,..,T
\end{aligned} \tag{6.21}$$

The change from the previous case without reservoir restriction is that the water values are now period specific, and that we have an explicit equation of motion for the reservoir-related shadow prices. Two consecutive water values are connected through the value of the shadow price on the reservoir constraint, as seen from the third condition in (6.21).

The reduced possibility of storing water may influence the strategy of importing and saving water for a higher price period. The possibility of overflow may restrict economically import of electricity since the water value may be driven down to zero in order to prevent overflow. In export periods the home price may be driven further up because there is a limit on the transfer of water from the previous period. If the reservoir constraint does not become binding we are back to the conditions in (6.12) for the situation without a reservoir constraint.

A bathtub illustration for two periods is provided in Figure 6.4. Since by design the foreign trade price is lowest in period 1 this period will be the import period. The figure is based on Figure 6.2. Inflow to the reservoir in period 1 is *AC* and in period 2 *CD*. The size of the reservoir is *BC*, indicated by $\overline{R}$, and the vertical broken lines from *B* and *C* represent the reservoir. The reservoir is introduced from *C* to the left to *B* because our dynamic problem for two periods is how much water to leave to period 2. The dotted left-hand wall of the hydro bathtub erected from *A* is extended to the left for period 1 indicated by the solid vertical axis line representing

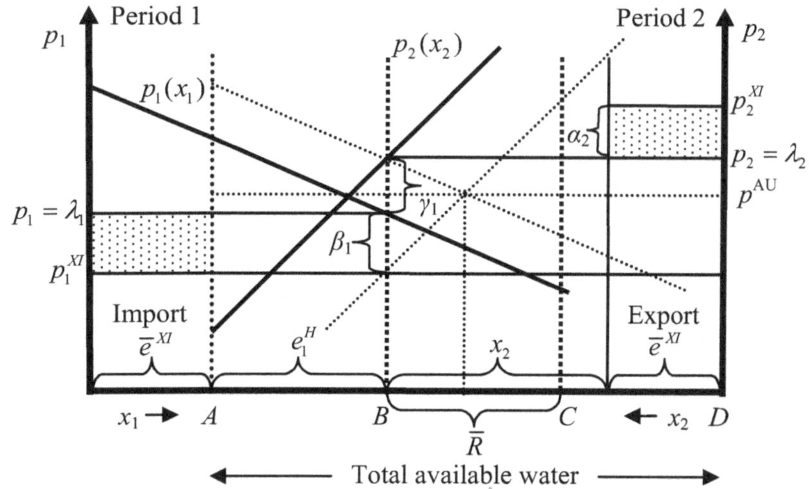

Figure 6.4. Reservoir and transmission limits.
Autarky indicated by dotted lines.

the import extension of the bathtub. In our case the full import capacity will be utilised. The full export capacity will also be used, and this capacity is indicated by the first solid line to the left of the right-hand hydro wall and to the right of point *C*. The way the figure is constructed trade is not extending the hydro bathtub wall in an export period to the right, but because export is at the expense of home consumption the new wall is erected to the left.

To show the change from autarky with water as the only resource and with a constrained reservoir to a situation with trade with a restriction, the final layout of the figure is the result of two stages for the two periods' curves. In the first autarky stage the demand curves indicated by dotted lines are anchored to the hydropower walls up from *A* and *D* (shown explicitly only for period 1). The dotted price and quantity allocation lines indicate the equilibrium situation for prices and allocation of electricity in autarky. The reservoir is not utilised to the upper constraint and the water values are equal and equal to the common social price p^{AU}. We then move on to the second stage with trade. For the import period we have that the whole capacity should be utilised. The demand curve for period 1 is shifted horizontally to the left and anchored on the import wall along the left-hand axis. Water *AB* will be used in period 1, and the import is maximal at $\bar{e}^{XI}$. The second optimality condition of (6.21) tells us that the social price, consistent with the sum of hydropower and import, is higher than the import price by the shadow price β_1 on the import capacity constraint. The maximal amount of water, $BC = \overline{R}$, is transferred to period 2.

Checking period 2 there is enough water to utilise the export capacity fully (all the available water ($BD - \overline{e}^{XI}$) will not be demanded for home consumption if the home price is set at the export price). The vertical solid line to the left of the right-hand hydropower wall then indicates the reduced availability for hydropower at home, and the demand curve is shifted horizontally with the distance of the export constraint to the left and anchored to this new wall (the actual anchoring point is not shown in the figure). The home price is found by the intersection of the demand curve and the hydropower wall for period 2 erected at B. According to the second optimality condition in (6.21) for export the home price is equal to the export price minus the shadow value on the export constraint. Since the export capacity is fully utilised the shadow price α_2 is positive and indicated as the difference between the export price and the home price for period 2 in the figure. The first condition in (6.21) tells us that the home price is equal to the water value.

The reservoir capacity has become constrained in the case with trade compared with autarky. The shadow price γ_1 on the reservoir capacity is found from the dynamic third condition in (6.21) and does in the figure indicate the difference between the two periods' water values.

Comparing the solution without trade and with restricted trade it is interesting to note that a situation where the reservoir is not used to its capacity and the period prices are equal, is turned into a situation where the reservoir is utilised maximally and the period prices are different. But the prices are not equal to the import- and export prices since both import and export is constrained, but lie between these two prices. The price in the import period becomes lower than the autarky price and the price in the export period becomes higher. The straightforward implication is then that electricity consumption in the home market in the import period 1 is higher than in the autarky solution, and the consumption is lower in the export period 2. The maximal amount is not transferred from period 1 to period 2 for the reason of enjoying higher consumption in period 2, but to give room for maximal export and earn money. Since trade volumes are equal the electricity trade is run at a surplus. Buying cheap and selling high is a classical principle for profitable trades. There is a congestion rent on the interconnector capacity in both periods indicated by the marked areas.

In the multiperiod case the strategy for reservoir accumulation and the possibility of processing maximal water and converting this into profitable export can result in a complicated pattern of import, accumulation of water, and releases for export earnings. The size of the reservoir compared with the maximal volume of exports will play a decisive role. The reservoirs can be managed without overflow because there is no production (or effect) restriction regarding home consumption, but the transmission constraint controls how much can be earned in high export price periods. Instead of having all water available for any period now, the accumulation of water by either holding back home consumption or by using import for home consumption instead of water is a more complicated strategy to follow.

Qualitatively the delimitation into the sets of export periods, import periods, and use of both hydro and import at home carries over from the previous section. The shadow-price dynamics expressed by the third condition in (6.21) does not influence qualitatively the classification of periods, but will, of course, influence the magnitudes involved. The trade prices will be the home prices whenever the trade restrictions do not bind.

Trade between countries hydro and thermal

So far we have operated with only a hydro economy. We will naturally term this country Hydro. We will now look at another country having only thermal capacity and it will be termed Thermal. The autarky situation and trade to fixed prices have been worked through for Hydro in the sections above, and we will now have a look at Thermal.

Trade with exogenous prices for a thermal economy

The properties of the capacity of Thermal and the aggregate merit-order variable cost function are described in Chapter 5. We will adopt problem (6.2) using the variable cost function (5.8) in Chapter 5 for aggregated thermal capacity. Looking at unrestricted trade, facing exogenous trading prices the social optimisation problem is:

$$\max \sum_{t=1}^{T} [\int_{z=0}^{x_t} p_t(z)\,dz + p_t^{XI} e_t^{XI} - c(e_t^{Th})]$$

subject to

$$x_t = e_t^{Th} - e_t^{XI} \tag{6.22}$$

$$e_t^{Th} \leq \bar{e}^{Th}$$

$$x_t, e_t^{Th} \geq 0,\ e_t^{XI} \text{ unrestricted}$$
$$T, p_t^{XI}, \bar{e}^{Th} \text{ given },\ t = 1,..,T$$

The symbols used for trade variables and their interpretations are the same as in the first section.

Eliminating the variable home consumption by substituting from the energy balance, the Lagrangian function is written:

$$L = \sum_{t=1}^{T} [\int_{z=0}^{e_t^{Th} - e_t^{XI}} p_t(z)\,dz + p_t^{XI} e_t^{XI} - c(e_t^{Th})] - \sum_{t=1}^{T} \theta_t (e_t^{Th} - \bar{e}^{Th}) \tag{6.23}$$

The necessary first-order conditions are:

$$\begin{aligned} &\frac{\partial L}{\partial e_t^{Th}} = p_t(e_t^{Th} - e_t^{XI}) - c'(e_t^{Th}) - \theta_t \leq 0\ (= 0 \text{ for } e_t^{Th} > 0) \\ &\frac{\partial L}{\partial e_t^{XI}} = -p_t(e_t^{Th} - e_t^{XI}) + p_t^{XI} = 0 \\ &\theta_t \geq 0\ (= 0 \text{ for } e_t^{Th} < \bar{e}^{Th}), \qquad t = 1,..,T \end{aligned} \tag{6.24}$$

First of all we note that the problem (6.22) is not a dynamic one under our assumptions. Each period can be solved in isolation, provided there are no restrictions on trade balance for a certain number of periods, e.g., for the time horizon T.

The second condition above tells us that with unrestricted trade the home price will always be set equal to the trade price of electricity. We assume that electricity is consumed in every period. If the marginal cost at zero output is higher than the trade price, nothing is produced at home and total consumption is imported. In export periods the first condition must hold with equality since power is generated in Thermal. If thermal capacity is fully utilised the shadow price on the capacity is switched on and added to the marginal cost.

The situation for two periods may be illustrated as in Figure 6.5 using two quadrants. Period 1 consumption is measured to the left of the central price- and marginal cost axis erected vertically from the origin O. Period 2 consumption is measured to the right. The marginal cost functions are identical and are drawn as straight lines upwards to the left and right from the common anchoring point at $c'(0)$ on the central axis. The short vertical lines at the end of the marginal cost curves indicate the limited capacity.

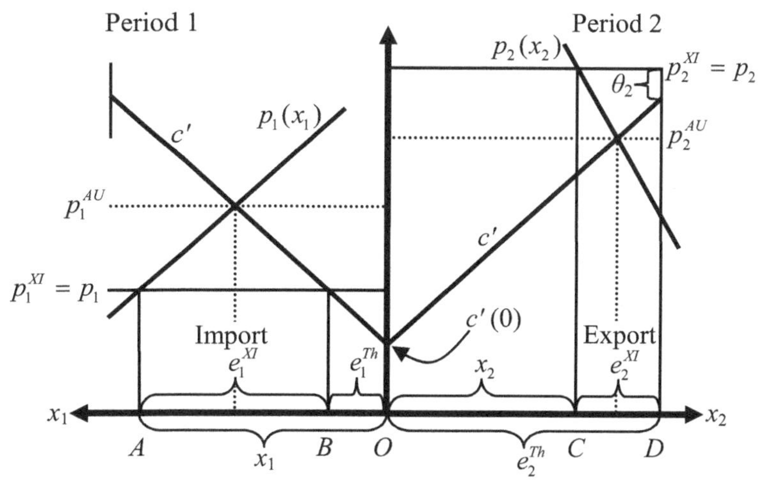

Figure 6.5. Thermal country and unconstrained trade.
Autarky indicated by dotted lines.

The demand curves are also straight lines for ease of exposition. Period 1 demand is made more elastic with a considerably lower choke price, resulting in a lower autarky price and quantity than the autarky situation for period 2. Period 1 may be called summer and period 2 winter. Introducing unlimited trade to the trade prices shown in the figure, with trade price for summer being lower than the autarky price (price line is shown by the dotted line) and vice versa for winter, it is optimal to import in summer the amount *AB* indicated in the second quadrant, but to export the amount *CD* in winter as shown in the first quadrant. Home production is undertaken only if it is cheaper than import, and export is undertaken if the trade price is higher than the autarky price. The figure illustrates that it may become profitable to expand the use of capacity in export periods right up to the capacity limit. In accordance with the first condition in (6.24) the shadow price on the capacity constraint, θ_2, is the difference between the trade price (equal to the home price) and the marginal cost at full capacity utilisation. The trading prices will be adapted as the home prices for Thermal. A production capacity constraint does not change this feature.

For the country Hydro, constraining the volume of trade in electricity provided some additional insights, but for Thermal it does not seem necessary because the thermal capacity is constrained in each period. The extreme result for Hydro without restriction on trade is due to the possibility of accumulating water over several periods and processing everything in the highest price period.

Trading with endogenous prices

So far we have operated with exogenous trade prices. But within an international market like the Nordic Nord Pool market equilibrium prices will be formed according to demand and supply. A stylised model with trade between the countries Hydro and Thermal will be explored. The opening up of trade between the neighbours Norway and Denmark has already been mentioned. Norway has a hydro share of 99%, and Denmark has a thermal share of 87% (2003). In a common market between Hydro and Thermal the production capacity of Thermal is given, and so is either the total amount of water within the planning horizon or reservoir capacity for Hydro. We will assume that Hydro and Thermal cooperate and are interested in a joint social solution. Value terms are expressed in the same money. Furthermore, any redistribution issues may be dealt with by side payments outside the electricity market.

In the electricity market with just the two countries, trade in electricity must balance in the sense that export from one country is the other county's import (and vice versa). The energy balance for each country can then be written:

$$\begin{aligned} x_t^H &= e_t^H + e_{Th,t}^{XI} - e_{H,t}^{XI} \\ x_t^{Th} &= e_t^{Th} + e_{H,t}^{XI} - e_{Th,t}^{XI} \end{aligned} \tag{6.25}$$

The quantities of electricity consumed in each country and exported, respectively imported, are now identified by country sub- and superscripts, "H" and "Th" for Hydro and Thermal, respectively. The superscript "XI" denotes export or import. When one country exports the other country cannot, but must import the identical volume (and vice versa).

The cooperative planning problem is first set up for the simplest case with a given amount of water at disposal (corresponding to assuming that the reservoir constraint will not become binding), and given thermal production capacity:

$$\max \sum_{t=1}^{T}\left[\int_{z=0}^{x_t^H} p_t^H(z)\,dz + \int_{z=0}^{x_t^{Th}} p_t^{Th}(z)\,dz - c(e_t^{Th})\right]$$

subject to

$$\begin{aligned} x_t^H &= e_t^H + e_{Th,t}^{XI} - e_{H,t}^{XI} \\ x_t^{Th} &= e_t^{Th} - e_{Th,t}^{XI} + e_{H,t}^{XI} \end{aligned} \tag{6.26}$$

$$\sum_{t=1}^{T} e_t^H \leq W$$
$$e_t^{Th} \leq \overline{e}^{Th}$$
$$x_t^H, e_t^H, e_{Th,t}^{XI}, e_{H,t}^{XI} \geq 0$$
$$T, W, \overline{e}^{Th} \text{ given } , t = 1,..,T$$

There is no restriction on the amount traded, but due to the way trade is set up for the two countries the traded amounts are non-negative.

In order to keep our problem as simple as possible, the country consumptions are substituted from the energy balances in the objective function when formulating the Lagrangian function:

$$\begin{aligned} L = \sum_{t=1}^{T} [& \int_{z=0}^{e_t^H + e_{Th,t}^{XI} - e_{H,t}^{XI}} p_t^H(z)dz + \int_{z=0}^{e_t^{Th} - e_{Th,t}^{XI} + e_{H,t}^{XI}} p_t^{Th}(z)dz - c(e_t^{Th})] \\ & - \sum_{t=1}^{T} \theta_t (e_t^{Th} - \overline{e}^{Th}) \\ & - \lambda (\sum_{t=1}^{T} e_t^H - W) \end{aligned} \tag{6.27}$$

The necessary first-order conditions are:

$$\begin{aligned} & \frac{\partial L}{\partial e_t^H} = p_t^H(x_t^H) - \lambda \leq 0 \ (= 0 \text{ for } e_t^H > 0) \\ & \frac{\partial L}{\partial e_{H,t}^{XI}} = -p_t^H(x_t^H) + p_t^{Th}(x_t^{Th}) \leq 0 \ (= 0 \text{ for } e_{H,t}^{XI} > 0) \\ & \frac{\partial L}{\partial e_t^{Th}} = p_t^{Th}(x_t^{Th}) - c'(e_t^{Th}) - \theta_t \leq 0 \ (= 0 \text{ for } e_t^{Th} > 0) \\ & \frac{\partial L}{\partial e_{Th,t}^{XI}} = p_t^H(x_t^H) - p_t^{Th}(x_t^{Th}) \leq 0 \ (= 0 \text{ for } e_{Th,t}^{XI} > 0) \\ & \lambda \geq 0 \ (= 0 \text{ for} \sum_{t=1}^{T} e_t^H < W) \\ & \theta_t \geq 0 \ (= 0 \text{ for } e_t^{Th} < \overline{e}^{Th}) \end{aligned} \tag{6.28}$$

Only conditions for export from a country (second and fourth) are entered because import is then determined residually. As the first step in a qualitative analysis of the optimal solution we assume that electricity is consumed in both countries in all periods. This implies that either hydro or thermal power has to be produced in every period. If Hydro is to export and Thermal import then the second constraint holds as an equality and the fourth constraint as an inequality with zero Thermal export, and vice versa. Whenever Hydro is exporting to Thermal the first condition in (6.28) implies that the market price in Hydro is equal to the shadow price on water. The second condition simply states the equilibrium condition that the domestic prices must be equal, and according to the first condition equal to the shadow price on water.

Thermal power is produced for home consumption in Thermal provided the marginal cost at zero production is less than the equilibrium price. The shadow price on the thermal capacity constraint is switched on if capacity is exhausted. If Thermal is exporting the third condition in (6.28) holds with equality and the second condition as an inequality with zero exports of hydropower.

Since the shadow price of water has a single value the equilibrium prices for all periods, where the situation described above is valid, become the same. If the shadow price of water should be higher than the home price in Hydro, then Hydro has to import from Thermal. This means that Thermal becomes the exporting country. The fourth condition tells us that the prices in the two countries must again be equal. Water is saved in such low price periods and used in the periods having a common, higher equilibrium price. When water is not used and Hydro is the importing country, equilibrium prices may vary. If water is used when Hydro is an importing country we are back to the regime with one single equilibrium price equal to the shadow price on water.

A two-period energy bathtub diagram may illustrate a possible optimal solution. Figure 6.6 is based on combining Figure 5.5 from Chapter 5 and Figure 6.5. The thin dotted lines all belong to the autarky situation marked in the figure with Hydro in the middle and Thermal as extensions at both sides. To the left and to the right of the hydro bathtub with floor *AB* and thin, solid wall-lines up from these points, the demand and supply curves for Thermal are entered for period 1 and 2, respectively. For period 1 on the left-hand side, demand and supply for Thermal is read from right to left, while the curves are read from left to right for period 2 on the right-hand side of the Hydro bathtub, as indicated in the figure. The marginal cost function is the same for both periods. The outer solid axes lines indicate the extensions of the hydro walls by full thermal capacity. The autarky price and quantity situation is indicated by thin dotted curves, while the curves relevant for the cooperative trade solution are drawn as solid lines.

To understand the figure better we may start with the autarky situation. The point *M* on the bathtub floor indicates the allocation of water on the

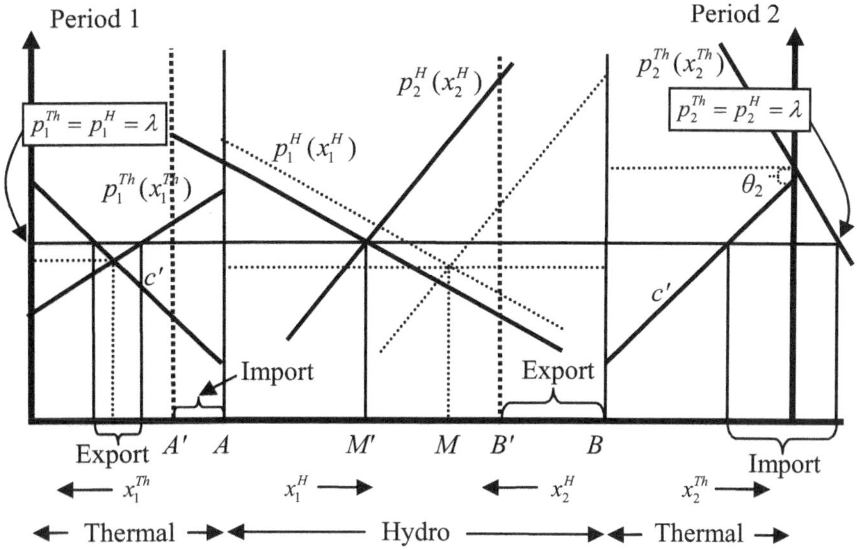

Figure 6.6. Trade between countries Hydro and Thermal. Autarky indicated by dotted lines.

two periods for Hydro. The prices are the same for both periods and determined by the intersection of the dotted demand curves. The common price is equal to the autarky shadow value of water. For Thermal the demand curves differ in such a way that while not all capacity is utilised in period 1, the whole capacity is used in period 2, resulting in the shadow price on the capacity constraint becoming positive. The period price becomes higher than marginal cost, as indicated in the figure on the right. This leads to a considerably higher electricity price in period 2 than in period 1 for Thermal in autarky, as also exhibited in Figure 6.5.

Now, introducing trade without restrictions on volumes the equilibrium solution is indicated by solidly drawn price and quantity lines. Since water is used in both periods in Hydro the prices for the periods become equal (remember that we have assumed the reservoir capacity limit not to become active). Furthermore, because the thermal capacity now is not exhausted in period 2 prices both across periods and countries become equal and equal to the shadow price on water, in accordance with the discussion of (6.28). The equilibrium price leads to Thermal exporting electricity in period 1 since the equilibrium price is higher than the autarky price. This export is then import to Hydro, and means that the wall erected from A gets a horizontal shift to the left to the vertical, broken line erected from A' with the amount of import. More electricity becomes available in Hydro.

The demand curve for period 1 also gets a similar horizontal shift and becomes anchored on the extended wall indicated by the broken vertical line erected from A'.

In period 2 Thermal reduces its production and gets reserve capacity again by substituting with imports from Hydro. The imports more than compensate for the reduction in thermal production. Since the export is at the expense of consumption in Hydro the solid bathtub wall originally erected from B gets a horizontal shift to the left with the length equal to the export. Less water is available for consumption in Hydro. The demand curve for period 2 gets a corresponding, horizontal shift to the left and is now anchored on the broken wall erected from B'.

Comparing autarky with trade we see that Thermal gets a higher price and a smaller volume in period 1 with trade, but the opposite is the case in period 2. Both the price reduction and volume increase are substantial. For Hydro somewhat less is consumed for the two periods seen together, leading to an increase in the price level in the trade regime. The allocation point on the bathtub floor $A'B'$ is M'. Notice that Hydro consumes, maybe surprisingly, less also in the import period. Water is stored in period 1 to be exported in period 2.

The extreme results with unrestricted trade that we saw for Hydro in the previous section studying the country in isolation are no longer the case. The fact that the prices are formed as equilibrium prices is enough to yield results that are plausible. However, we saw that Thermal gets an import in period 2 resulting in total consumption by far exceeding total production capacity in Thermal. It may be unrealistic that the transmission system in Thermal has a capacity to handle much higher volumes than it can generate itself. In addition it is of interest to see if a constraint on the reservoir induces other results concerning prices and quantities.

Trade with constraints on reservoir and trade volumes

Introducing constraints on reservoir and volume of trade the objective function (6.23) for the cooperative optimisation problem becomes:

$$\max \sum_{t=1}^{T} [\int_{z=0}^{x_t^H} p_t^H(z)dz + \int_{z=0}^{x_t^{Th}} p_t^{Th}(z)dz - c(e_t^{Th})]$$

subject to

$$x_t^H = e_t^H + e_{Th,t}^{XI} - e_{H,t}^{XI}$$

$$x_t^{Th} = e_t^{Th} - e_{Th,t}^{XI} + e_{H,t}^{XI}$$

$$
\begin{aligned}
& R_t \le R_{t-1} + w_t - e_t^H \\
& R_t \le \overline{R} \\
& e_{H,t}^{XI} \le \overline{e}^{XI},\ e_{Th,t}^{XI} \le \overline{e}^{XI} \\
& e_t^{Th} \le \overline{e}^{Th} \\
& x_t^H, x_t^{Th}, e_t^H, e_t^{Th}, e_{Th,t}^{XI}, e_{H,t}^{XI} \ge 0 \\
& T, w_t, R_o, \overline{R}, \overline{e}^{XI}, \overline{e}^{Th} \text{ given, } R_T \text{ free },\ t = 1,..,T
\end{aligned}
\tag{6.29}
$$

The two restrictions involving the reservoir do substitute for the total water constraint in (6.26). In addition to the restrictions on trade we could also consider restriction on hydropower production and on country transmissions, especially relevant for Thermal since we saw that consumption became higher than production capacity due to trade in Figure 6.5. Production is already constrained there. It is straightforward to introduce such constraints. We leave to the reader to introduce them, since it becomes too complicated to make a visually pleasing figure illustrating all constraints if they are to bind. As to transmission-capacity constraint it has to be linked to the consumption in Thermal, $x_t^{Th} \le \overline{x}^{Th}$, and similarly for Hydro. However, as in earlier chapters internal transmission capacity is disregarded and we focus on interconnector capacity between the countries.

Substituting for country consumptions from the energy balances in the objective function, the Lagrangian is:

$$
\begin{aligned}
L = & \sum_{t=1}^{T} [\int_{z=0}^{e_t^H + e_{Th,t}^{XI} - e_{H,t}^{XI}} p_t^H(z)dz + \int_{z=0}^{e_t^{Th} - e_{Th,t}^{XI} + e_{H,t}^{XI}} p_t^{Th}(z)dz - c(e_t^{Th})] \\
& - \sum_{t=1}^{T} \lambda_t (R_t - R_{t-1} - w_t + e_t^H) \\
& - \sum_{t=1}^{T} \gamma_t (R_t - \overline{R}) \\
& - \sum_{t=1}^{T} \alpha_{h,t} (e_{H,t}^{XI} - \overline{e}^{XI}) \\
& - \sum_{t=1}^{T} \alpha_{Th,t} (e_{Th,t}^{XI} - \overline{e}^{XI}) \\
& - \sum_{t=1}^{T} \theta_t (e_t^{Th} - \overline{e}^{Th})
\end{aligned}
\tag{6.30}
$$

The first-order necessary conditions are:

$$
\begin{aligned}
&\frac{\partial L}{\partial e_t^H} = p_t^H(x_t^H) - \lambda_t \le 0 \quad (=0 \text{ for } e_t^H > 0) \\
&\frac{\partial L}{\partial e_{H,t}^{XI}} = -p_t^H(x_t^H) + p_t^{Th}(x_t^{Th}) - \alpha_{H,t} \le 0 \ (=0 \text{ for } e_{H,t}^{XI} > 0) \\
&\frac{\partial L}{\partial R_t} = -\lambda_t + \lambda_{t+1} - \gamma_t \le 0 \ (=0 \text{ for } R_t > 0) \\
&\frac{\partial L}{\partial e_t^{Th}} = p_t^{Th}(x_t^{Th}) - c'(e_t^{Th}) - \theta_t \le 0 \ (=0 \text{ for } e_t^{Th} > 0) \\
&\frac{\partial L}{\partial e_{Th,t}^{XI}} = p_t^H(x_t^H) - p_t^{Th}(x_t^{Th}) - \alpha_{Th,t} \le 0 \ (=0 \text{ for } e_{Th,t}^{XI} > 0) \\
&\lambda_t \ge 0 \ (=0 \text{ for } e_t^H > 0) \\
&\gamma_t \ge 0 \ (=0 \text{ for } R_t < \bar{R}) \\
&\theta_t \ge 0 \ (=0 \text{ for } e_t^{Th} < \bar{e}^{Th}) \\
&\alpha_{H,t} \ge 0 \ (=0 \text{ for } e_{H,t}^{XI} < \bar{e}^{XI}) \\
&\alpha_{Th,t} \ge 0 \ (=0 \text{ for } e_{Th,t}^{XI} < \bar{e}^{XI}), \quad t = 1,..,T
\end{aligned}
\tag{6.31}
$$

Whenever reservoir constraints are involved we get a time-specific water value as shown in the first condition in (6.31), and an equation of motion for the reservoir shadow prices, here the third condition. If hydropower is produced the first condition holds with equality, and the period price in Hydro is equal to the water value. Furthermore, if hydropower is exported we have from the second condition that the social prices in the countries must be the common equilibrium price as long as the export capacity is not constrained, because according to the complementary slackness condition, the shadow price is zero. If hydropower export is zero, then the shadow price on the export-of-hydropower constraint is still zero. According to the second condition in (6.31) the prices in Hydro and Thermal may then differ, with thermal price being less than or equal to the hydropower price. The question is if such a difference can be part of a social solution in our model. With a lower thermal price the objective function could be increased by transferring a unit of thermal production to Hydro, i.e., exporting thermal power. But looking at the fifth condition for thermal export when it is positive, we have that the prices again have to be equal.

If the capacity constraint in Thermal is not binding, then the common equilibrium price that was established to be equal the equilibrium price, is also equal to the marginal production cost in Thermal.

If trade constraints are binding, both export and import will be binding for the same period. The second and fifth conditions in (6.31a) tell us that in such a situation it may be optimal to have different prices between the countries. The price will be lower in the country that is export-constrained than in the country that is import-constrained. An active export constraint forces the country to use more electricity at home, and to realise this, the price has to decrease. For an importing country the home price has to increase as a response to being rationed on imports.

Combining Figures 6.4 and 6.6, the impact of a reservoir constraint can be illustrated for two periods as in Figure 6.7. Hydro is described by a hydro bathtub in the middle extended by thermal capacity on each side. The bathtub floor is *AD*, and available water in period 1 is *AC* and *CD* in period 2. The amount *BC* can be stored in period 1 and transferred to period 2.

The dotted demand curves and the hydro bathtub walls with solid vertical lines erected from *A* and *D* show the autarky solution for Hydro. The autarky solution for Thermal is similar to the solution shown in Figure 6.5. The country-specific equilibrium in price and quantities are indicated by

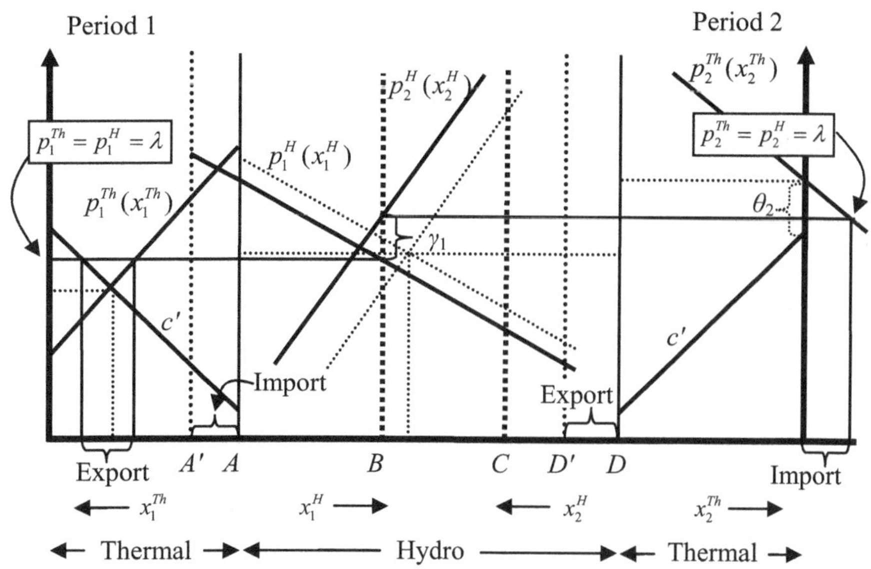

Figure 6.7. Trade between Hydro and Thermal with a reservoir constraint. Autarky indicated by dotted lines.

the dotted lines. We have that for Hydro the autarky prices are equal for the periods. The reservoir capacity *BC* is not fully utilised in Hydro transferring water from period 1 to period 2 to obtain the social autarky solution. The period 1 price for Thermal is lower than in Hydro, while the period 2 price is higher. The capacity in Thermal is constrained in period 2, and a shadow price is switched on to keep demand within the limits set by autarky supply at maximal capacity.

Opening up for trade we have a common equilibrium price forming for period 1 just as explained for Figure 6.6. The bathtub wall for period 1 for Hydro gets a horizontal shift to the left, indicated by the dotted vertical line erected from *A'*, equal to the import to Hydro in period 1. The equilibrium price is just slightly lower than the autarky price. What is remarkable is that the water use is changed markedly between the two periods compared with autarky. Now a full reservoir *BC* is transferred to period 2. Since the equilibrium price is slightly lower in period 1 with trade the total electricity consumption is also a little greater. But notice that the use of water in period 1 goes down.

The autarky price for Thermal in period 2 suggested export possibilities for Hydro since the Hydro autarky price was considerably lower. A maximal amount in the reservoir is now saved for use in period 2. The common equilibrium price in period 2 is found after shifting, for Hydro, the demand curve and bathtub wall from the right-hand bathtub wall erected from *D* to the left, indicated by the dotted vertical line from *D'*, with the horizontal shift being equal to the export of hydropower to Thermal. Then the price is determined by the intersection of the shifted demand curve and the broken line erected from *B* representing the maximal reservoir and the start of water available for period 2. The difference in prices between the two periods is expressed by the shadow price γ_1 on the upper reservoir constraint. The price in period 2 in Thermal does not decrease sufficiently for spare generating capacity to develop. The capacity is still constrained, but the shadow price on this constraint is considerably less, indicating a long-term benefit for Thermal since building out more capacity may be postponed. For Hydro we note that the equilibrium price is higher than the autarky price, leading to lower electricity consumption with trade, i.e., less water is used at home due to export.

The trade benefits Thermal in period 2 with lower price and higher consumption compared with autarky. In period 1 the pattern is reversed. Since the trades are almost equal Thermal gets a deficit on the electricity trade, and Hydro a corresponding surplus since the equilibrium price is lower when Thermal exports than when it imports, and vice versa for Hydro.

We dropped the constraints on production and internal country transmission capacity in the model above. We can use Figure 6.7 to indicate possible influences of such constraints when they are binding. If Thermal has a domestic transmission network constraint that does not allow the full consumption in period 2 as shown, then the constraint will force a lower consumption, lower import, and a higher price in period 2. The prices will now differ between the countries in period 2. Hydro will export less. The motivation for storing maximal water in period 1 is weakened and the constraint may lead to the reservoir storage not being completely filled. The implication is that Hydro may consume more water in both periods; the equilibrium price in period 1 will decrease and reduce the export from Thermal and increase consumption.

The day trade between Norway and Denmark is often mentioned as an example of gains by trade when hydropower with storage is coupled with a thermal system. Norway can import thermal power in the night time and accumulate water in the reservoirs when demand in both countries is low (see Figure 1.2 in Chapter 1 for demand variation over 24 hours in Norway) and only the most cost-efficient thermal plants are generating power, and then export hydropower in daytime and save Denmark for taking into use the least cost efficient thermal plants. If we think about one hour as the period definition in model (6.26), Figure 6.6 may illustrate this development of trade over day and night. If period 1 represents night time and period 2 daytime, then we just have export from Thermal during the night and import to Hydro, accumulating more water than in autarky, and the reverse in daytime: export of hydro and import to Thermal. The two flows are about equal, but the flows may, of course, differ in real life. Since more capacity is used in Thermal in night time the marginal cost is pushed up, but there is no reduction in marginal costs during daytime in Thermal because in our example the capacity is also exhausted in that period. The capacity utilisation increases in Thermal.

Chapter 7. Transmission

So far the transmission system has not been modelled, although it is a physical necessity to have a network. The main reason was that the existence of a network does not play an explicit role for the dynamics of the hydropower system. Assuming network capacities to be given, the flow of electricity is continuous and does not influence the nature of the dynamic equations driving an optimal plan over periods. However, network effects may influence the quantitative solutions in a way that is different for a hydropower system than systems with, e.g., thermal generation. It may also be interesting to consider transmission regarding the spatial structure of pricing of electricity based on hydro generation, since hydro can be almost instantaneously switched on and off with modest costs. Hydropower may therefore be the most suitable generating technology for applying spatial pricing. One key question is whether transmission as a service should be priced separately by a social planner, and whether such transmission costs may influence the time profiles of utilising reservoirs. There is also the issue of impact of limited capacities on network lines and the price structure. We will investigate changes in our model analyses implied by networks, and especially look for impacts on use of hydropower.

Making transmission explicit we have as a basic unavoidable feature that some electricity is lost in the network because that the current of electricity through conductors creates heat. The average loss is in the range of 2-3% in high-voltage transmission in national or regional grids, and 5-15% in low-voltage distribution networks supplying the residential sector and other low-voltage customers (220-240 volts in Europe, 110-120 volts in the United States). However, these losses are average values, and marginal values may be considerably higher (in general loss is a function of the square of the energy flow).

Transmission is governed by physical laws like Ohm's law and Kirchhoff's laws securing lowest possible loss in a network system of generating nodes and consuming nodes,[1] given what is put in and what is

[1] According to Bohn et al. (1984) this version of Kirchhoff's laws works for direct current, although they have not been able to prove it for alternating current; however, it is a useful way to think about electricity flows.

taken out. Changing spatial supply and consumption configurations may change the loss and consumption for given total supply. Complicated physical laws (at least for economists) are involved. Based on concepts like electrical angles and reactive power, patterns of flows may change rapidly and total energy delivered both be reduced and even increased by more than the increase in input. Pursuing this takes us outside the scope of the present book, so we will only point to such effects and model transmission in a way that make some of these effects possible to unfold [see Schweppe et al. (1988) for extensive elaborations based on physical laws and the restatement in Hsu (1997) of the main points of transmission modelling there].

Engineering approach to transmission in economics

The transmission of electricity is a classical example in economics of an engineering production function (Førsund, 1999). According to Vernon L. Smith the problem of finding the cost efficient way of setting up a transmission line between a node with electricity generation and a node with consumption was first analysed by Lord Kelvin (William Thomson) in 1881 (Thomson, 1881). The two-node model is illustrated in Figure 7.1.

The basic laws governing electrical flows used by Smith (1961, pp. 24-30) deriving the engineering production function for transmission are the following:

$$\begin{aligned} P_o &= P_i - P_L \\ P_L &= I^2 R \\ P_o &= V_o I \cos\varphi \Rightarrow I = \frac{P_o}{V_o \cos\varphi} \end{aligned} \tag{7.1}$$

where the definitions of the variables are:

P_o = the consumption of power in kW
P_i = the generation of power in kW
P_L = the loss in kW
I = current in amps
R = resistance of the line in ohms
V_o = fixed voltage at the consumer node
$\cos\varphi$ = power factor of the consumer's load
φ = lag between voltage and current variation in an alternating-current circuit.

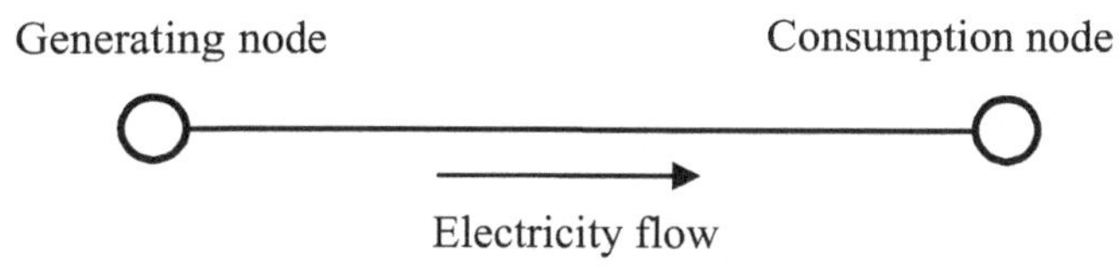

Figure 7.1. A network with one generating node and one consumption node

The first equation states the conservation of energy, i.e., that the power received by the consumer is the difference between the generation of power and the loss in the line due to resistance. The second equation is Ohm's law and the third equation expresses the definitional connection between power, voltage, and current.

Ohm's resistance is related to the length of the line, $2L$, (L is the length between the generating node and the consumer node) the specific resistance of the type of metal used, ρ, and the cross-sectional area, A, of the cable:

$$R = \frac{2L\rho}{A} \tag{7.2}$$

Substituting from the last line in (7.1) and from (7.2) into the first equation in (7.1) yields:

$$P_o = P_i - P_L = P_i - I^2 R = P_i - \left(\frac{P_o}{V_o \cos\varphi} \right)^2 \frac{2L\rho}{A} \tag{7.3}$$

Introducing the weight of the cable, K, we have that $K = 2dLA$, where d is the specific weight of the metal used.

Finally, Smith (1961) derived the following long-run transformation function on implicit form between electricity received, renamed x (kW), as output and electricity generated, renamed e (kW), and weight of the conductor, K, as inputs by multiplying the terms in (7.3) with K:

$$F(x,e,K) = -K(e-x) + kx^2 = 0\ ,\ k = \frac{4L^2 d\rho}{(V_o \cos\varphi)^2} \tag{7.4}$$

We have used the standard convention that partial derivatives of the transformation function with respect to inputs are negative, and that the partial derivative with respect to output is positive. The constant k sums up the engineering information necessary for the parameterisation of the production function. The constant k will depend on the type of metal chosen for the conductor through specific weight and resistance. The difference $(e-x)$ is the power lost due to resistance.

We have used energy and not power as the dimension of our variables previously. It is straightforward in principle to convert the power variables in (7.4) into energy during the time period in question by either integrating over continuous time within the period, or using discrete time and the average loads within each time interval and multiplying. For short enough time periods an assumption of constant power in continuous time may be used as an approximation.

In order to facilitate the exhibition of substitution- and scale properties the transformation function (7.4) can be solved explicitly (not done in Smith, 1961) for output as function of inputs and parameters:

$$x = f(e,K) = \frac{K}{2k}\left[(1+4\frac{ke}{K})^{\frac{1}{2}} - 1\right] \tag{7.5}$$

The marginal productivity of e is positive and decreasing:

$$\frac{\partial x}{\partial e} = (1+\frac{4ke}{K})^{-1/2} > 0, \ \frac{\partial^2 x}{\partial e^2} = -\frac{2k}{K}(1+\frac{4ke}{K})^{-3/2} < 0 \tag{7.6}$$

This long-run production function exhibits constant returns to scale in electricity input and weight of conductor; multiplying e and K with a scalar quantity yields that the output is also multiplied with this quantity.

The *ex ante* rate of substitution between the weight of the conductor and energy generated at the production node can most easily be worked out using the implicit form (7.4):

$$MRS = -\frac{dK}{de} = \frac{K}{e-x} > 0 \tag{7.7}$$

The sign is correct since the denominator must be positive. Reducing the weight of a conductor of a specific metal increases the energy needed to be generated in order for the consumers to receive a certain amount of electricity at a given voltage.

When an input in a production function that exhibits constant returns to scale is kept fixed, we know that for the remaining variable inputs the returns to scale must be less than one. In the short run when the conductor is capital in place and fixed, the production function (7.5) exhibits diminishing returns to the remaining inputs, i.e., the electricity input. This can be worked out using the marginal productivity of electricity input in (7.6). There is diseconomy of scale in the short run. Keeping the physical conductor constant, increasing injection at the generation node with 1% increases the energy reaching the consumer node with less than 1%.

The flow of electricity through the line obviously has an upper physical limit that we do not model using the production function (7.5). There is a design limit to the amount of current the line can carry without being damaged by the heat created due to resistance.

The problem stated by Lord Kelvin in 1881 was to find the conductor (represented by the area of the cross-section) minimising costs. Rephrasing the problem as one of minimising *annualised* costs, using the transformation function (7.4) and introducing p_e as the fixed price of electricity input (this price may be linked to marginal generating costs), p_K as the fixed price per unit of weight (for a given length) of the conductor and r for the capital annualisation factor (equal to the rate of discount for an infinite length of life of the conductor), the formal problem is:

$$\begin{aligned}
&\min C = p_e e + r p_K K \\
&\text{subject to} \\
&x = x^o \\
&-K(e-x) + kx^2 = 0 \Rightarrow \\
&e = x^o + \frac{kx^{o2}}{K}
\end{aligned} \tag{7.8}$$

The given output level x^o is assumed to be below the capacity of the line. Substituting for energy input using the last equation above, and setting the partial derivative of the resulting cost expression with respect to K equal to zero, yields the condition for the weight of a specific choice of conductor:

$$\begin{aligned}
-p_e \frac{kx^2}{K^2} + r p_K = 0 \quad &\Rightarrow r p_K K = p_e \frac{kx^2}{K} \\
&\Rightarrow r p_K K = p_e (e-x) = p_e I^2 R \, ,
\end{aligned} \tag{7.9}$$

where the transformation function is used in the last line, and the last expression follows from Ohm's law (the superscript for the given output level is dropped). For simplicity, following the original discussion, direct current is considered (or the perfect condition of an AC-system of $\varphi = 0$ is assumed). We see that loss I^2R is proportional to the square of the current. For a specific type of conductor the diameter or weight should be chosen such that the current value of the loss created by transmission is equal to the annualised cost of the conductor measured by weight. Inspecting a set of feasible conductors, the type of metal implying the lowest cost of loss, given (7.9), should be chosen.

The production function (7.5) is too simple to portray real transmission, and leave out, e.g., economies obtained by reducing losses by transmitting electricity at high voltage. At each end of our stylised transmission there are transformers bringing the voltage up for transmission and bringing it down again at consumer nodes to the appropriate voltage for consumers. The practicalities of weighing high voltage and accompanying smaller loss against need for transforming the voltage result in a level of transmission networks of highest voltage for the national grid, then a level of less voltage at regional networks and a level of lowest voltage for distribution networks within consumption nodes.

Modelling transmission for simple cases

In order to capture the essence of the transmission activity, i.e., the spatial aspect, multiple generating plants must be specified. It is usual to call points in a network where generation and consumption take place, for *nodes* (or buses). There may be one or more generators within a node, and one or more types of end-consumer (households, firms, agriculture, etc.). In a hydropower system the location of generation is determined by natural conditions, and the overlap between generation nodes and consumer nodes may not be that great. When transmission is introduced a new endogenous variable is also necessarily introduced; the loss incurred by heat created in the conductors when electricity flows through the networks. As mentioned introductory it will be of specific interest in this book with the focus on hydropower whether introducing transmission brings in new dimensions as to the utilisation of hydropower plants, both across space within the same period, and over time.

Two nodes and two periods

We will start by first specifying only one consumer node with an aggregate demand function and one generation node, as portrayed in Figure 7.1, in order to bring out the basic new features when transmission is introduced. The generation of electricity is done using hydropower. The question is whether the introduction of transmission will have any impact on how hydropower is utilised over time. Only two periods will be considered first in order to keep the analysis as simple as possible, thus making it possible to adapt a bathtub diagram for illustration.

The new features to include in the model of the type studied in Chapter 3 concerns the energy balance telling us that energy consumption is equal to energy generated subtracted the loss on the line between the generator and the consumer, as in the first equation in (7.1). Using our standard symbols for consumption and production, and introducing e_t^L for loss, the energy balance is

$$x_t + e_t^L = e_t^H, \; t = 1,..,T \tag{7.10}$$

Our variables are now measured as in earlier chapters in energy units (kWh). In order to capture the physical laws expressed by the two last equations in (7.1) we just state that the loss (in kWh) created within a period is a function of the energy received at the consumption node, keeping in mind the transformation from power concepts to energy concepts as explained in the previous section:

$$e_t^L = e_t^L(x_t), \frac{\partial e_t^L(x_t)}{\partial x_t} > 0, \frac{\partial^2 e_t^L(x_t)}{\partial x_t^2} > 0, \; t = 1,..,T \tag{7.11}$$

We only need that the first- and second-order derivatives are positive for qualitative analyses, but more specific expressions may be worked out using Ohm's law as shown in (7.1). The signing is based on Ohm's law.

Capacity limits on lines are important for how a transmission system behaves. We will assume a unique relationship between the physical limit on how much heat, created due to resistance, a line can safely be exposed to, and the limit on energy delivered to the consumer node. This is in line with our earlier discussion of going from power variables in kW to energy variables in kWh. In our analysis a situation with a binding line constraint is called congestion. However, in reality the situation is not so "zero – one", since it takes some time before excessive heat makes permanent damage to a line, making the line sag or eventually break.

In order to focus on the aspects of transmission we will only use the reservoir constraints and not introduce the other constraints listed in Table 3.1 in Chapter 3. The social planning problem for one generating node, one consumer node, transmission between the nodes, and two time periods can then be set up as follows:

$$\max \sum_{t=1}^{2} \int_{z=0}^{x_t} p_t(z)dz$$

subject to

$$R_t \le R_{t-1} + w_t - e_t^H$$

$$R_t \le \overline{R}$$

$$
\begin{aligned}
& x_t + e_t^L = e_t^H \\
& e_t^L = e_t^L(x_t) \\
& x_t \le \overline{x} \\
& x_t, e_t^H, e_t^L \ge 0,\ t = 1,2 \\
& R_o, \overline{R}, \overline{x} \text{ given}
\end{aligned}
\tag{7.12}
$$

The first two restrictions concern the reservoir dynamics and capacity constraint. The third equality restriction is the energy balance for period t expressing that consumption and loss add up to generation. Loss on a line is created in a complex way physically, but here it boils down to the loss being a function of the amount of consumption. The restriction on how much power the line can carry within safety standards is also related to the consumption, remembering that the underlying assumption of using energy as a variable for a time period is that the power level is constant in continuous time within the interval.

Eliminating the loss variable, the Lagrangian for the highly stylised problem is:

$$
\begin{aligned}
L = & \sum_{t=1}^{2} \int_{z=0}^{x_t} p_t(z)dz \\
& -\sum_{t=1}^{2} \lambda_t (R_t - R_{t-1} - w_t + e_t^H) \\
& -\sum_{t=1}^{2} \gamma_t (R_t - \overline{R}) \\
& -\sum_{t=1}^{2} \tau_t (x_t + e_t^L(x_t) - e_t^H) \\
& -\sum_{t=1}^{2} \mu_t (x_t - \overline{x})
\end{aligned}
\tag{7.13}
$$

The Lagrangian parameter τ_t for the energy balance is free in sign because the energy balance must hold with equality.

The necessary first-order conditions are:

$$
\begin{aligned}
& \frac{\partial L}{\partial x_t} = p_t(x_t) - \tau_t - \tau_t \frac{\partial e_t^L}{\partial x_t} - \mu_t \le 0 \ (= 0 \text{ for } x_t > 0) \\
& \frac{\partial L}{\partial e_t^H} = -\lambda_t + \tau_t \le 0 \ (= 0 \text{ for } e_t^H > 0)
\end{aligned}
$$

$$\frac{\partial L}{\partial R_t} = -\lambda_t + \lambda_{t+1} - \gamma_t \leq 0 \ (= 0 \text{ for } R_t > 0) \tag{7.14}$$
$$\lambda_t \geq 0 (= 0 \text{ for } R_t < R_{t-1} + w_t - e_t^H)$$
$$\gamma_t \geq 0 (= 0 \text{ for } R_t < \bar{R})$$
$$\mu_t \geq 0 (= 0 \text{ for } x_t < \bar{x}) \ , t = 1,2$$

We will assume that electricity is consumed and produced in both periods so the first two first-order conditions hold with equality. The second condition tells us that the shadow price on the energy balance is equal to or less than the non-negative shadow price on the water accumulation constraint; the water value. In the case of overflow the water value becomes zero, and then so will the shadow price on the energy-balance constraint when production is positive. The social price will then also become zero unless the upper constraint on the line is reached.

It may be the case that line capacity is so restricted that not all available water can be utilised. For this to happen the line constraint must be binding in both periods. The social price is then determined only by the shadow price on the line capacity, since the water values will be equal to the shadow price on the energy constraint and equal to zero. Water is left in the reservoir at the end of period 2. The transmission constraint leads to a lock-in of water.

However, it seems more reasonable to assume that the line is dimensioned in such a way that water is not lost. The condition is that the sum of available water over the two periods is less than twice the upper capacity on the line; $R_0 + w_1 + w_2 < 2\bar{x}$. Assuming interior solutions the water value is positive and then the shadow price on the energy balance is equal to the period's water value. The social price for a period is in this case positive and composed of the shadow price on water plus the value of the marginal loss generated. The loss is valued using the common water value and shadow price on the energy-balance constraint. The water value represents marginal production cost in the form of an opportunity cost. The water value will vary between the time periods if the reservoir constraint becomes binding.

The difference between the social price at the consumer node and the water value at the production node is made up of the marginal loss and congestion terms:

$$p_t(x_t) - \lambda_t = \lambda_t \frac{\partial e_t^L}{\partial x_t} + \mu_t \ , t = 1,2 \tag{7.15a}$$

In the case of congestion the shadow price on the line capacity constraint is added to the loss term. The social price will be greater than the water value when there are losses and/or congestion.

The social prices may now become different between the two periods due to the loss and congestion terms:

$$p_2(x_2) - p_1(x_1) = \lambda_2 \frac{\partial e_2^L}{\partial x_2} - \lambda_1 \frac{\partial e_1^L}{\partial x_1} + (\mu_2 - \mu_1) \tag{7.16a}$$

The relative size of the loss and congestion terms of the two periods will determine which period price is the highest. Since we have just one line only one congestion term may be positive in (7.16a) for the period with the highest consumption.

The third equation in (7.14) is the equation of motion for the water values. Transmission-related variables are not explicitly appearing, but we will study how transmission can influence the running of hydropower for the two periods. Let us first assume that the upper reservoir constraint will not be reached in the first period, and that the reservoir is emptied in the last period, and that the social price remains positive. The dynamics of the water-related shadow prices then tell us that the water value will be the same for the two periods. This implies that the shadow price on the energy balance will also be the same for both periods and equal to the common water value. Equations (7.15a) and (7.16a) can then be rewritten

$$p_t(x_t) - \lambda = \lambda \frac{\partial e_t^L}{\partial x_t} + \mu_t \ , t = 1,2 \tag{7.15b}$$

$$p_2(x_2) - p_1(x_1) = \lambda\left(\frac{\partial e_2^L}{\partial x_2} - \frac{\partial e_1^L}{\partial x_1}\right) + (\mu_2 - \mu_1) \tag{7.16b}$$

A bathtub diagram may illustrate the situation, first dropping congestion effects for simplicity. In Figure 7.2 the demand in period 1 is lower than the demand in period 2 for all price levels. The demand curves and the curves $(p_t(x_t) - \lambda\partial e_t^L/\partial x_t)$, $t = 1,2$, are assumed to be linear, i.e., the change in the marginal loss is assumed to be constant for the latter curve.

This is, in fact, in accordance with Ohm's law saying that the marginal loss is twice the average loss. The bathtub floor is the total available water, $R_o + w_1 + w_2$. The amount AC is available in the first period, and the inflow in period 2 is CD. However, the erection of the bathtub walls must now reflect the losses created in the two periods, so the walls start on the inside of the availability line on both sides. The placement of the walls is determined endogenously as a solution to the model (7.12) above. The optimal

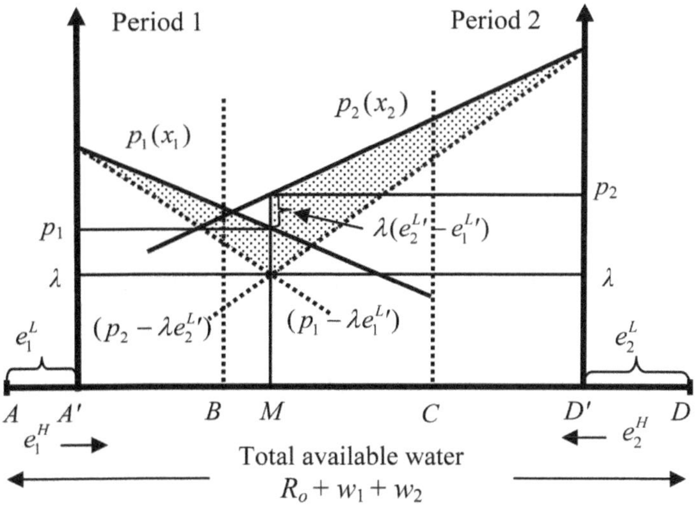

Figure 7.2. Impact of network loss on social prices

solution is found, using (7.15b), at the intersection of the curves ($p_t(x_t) - \lambda\partial e_t^L/\partial x_t$) for the two time periods determining the level of the common water value.[2] The reservoir capacity is *BC*, and we see that the reservoir capacity is not constrained in the optimal solution. The social prices are found by going up to the respective demand curves. *A'M* is consumed in the first period, *MC* is transferred from period 1 to period 2, and *MD'* is consumed in period 2. As long as the demand curves differ between the two periods the price will be higher for the period with the highest demand due to the greater loss generated. The social price is higher in the high-demand period in order to discourage consumption in the high-demand period when losses are also considered in the optimisation. This happens although the water value is the same for both periods. We have found a new reason for price differences in a pure hydropower system.

The losses are illustrated in an ad hoc way as *AA'* and *D'D* with *D'D* > *AA'*. However, the value of the losses can be identified in the figure as shown by the shaded triangles.

In the case of a binding reservoir constraint the situation is qualitatively different. It turns out that the difference between the social prices is now determined by the reservoir constraint as analysed in Chapter 3. However, the absolute effects are influenced by the losses created. Figure 7.3 illustrates the situation. The reservoir constraint is binding imposing a limit on

[2] In Figures 7.2, 7.3 and 7.4 the partial derivative $\partial e_t^L / \partial x_t$ is written $e_t^{L\prime}$.

the transfer of water to period 2. From the third condition in (7.14) we have that the shadow price on the reservoir constraint in the first period becomes positive. The water values will therefore differ, with $\lambda_2 = \lambda_1 + \gamma_1$.

When calculating the value of the loss in Equation (7.15a) the calculation price for period 2 is then greater than the calculation price for period 1. This is reflected in the relative size of the gap between the demand curves and the marginal loss curves in the figure. The allocation of water between the periods is now determined by the size of the reservoir since a full reservoir is transferred to period 2. But notice that since the bathtub walls are endogenously erected transmission losses are indirectly influencing the absolute allocation. In fact, restricting the amount that can be transferred to period 2 will increase the use of water in period 1 and decrease it in period 2, leading to somewhat smaller total losses, assuming that the consumption in period 2 is greater than consumption in period 1. This is indicated by the relative size of the losses in the figure. The loss in period 2 is still greater than the loss in period 1.

The consumer prices are determined by the intersections between the demand curves and the vertical broken line for the reservoir capacity from *B*. But the difference between the prices is no longer the shadow price on the reservoir constraint as in Chapter 3, but is expressed by Equation (7.16a). Eliminating the water value for period 1, we see that the price difference is an expression involving differences of marginal losses for the two periods, evaluated using the water value for period 2, and the evalu-

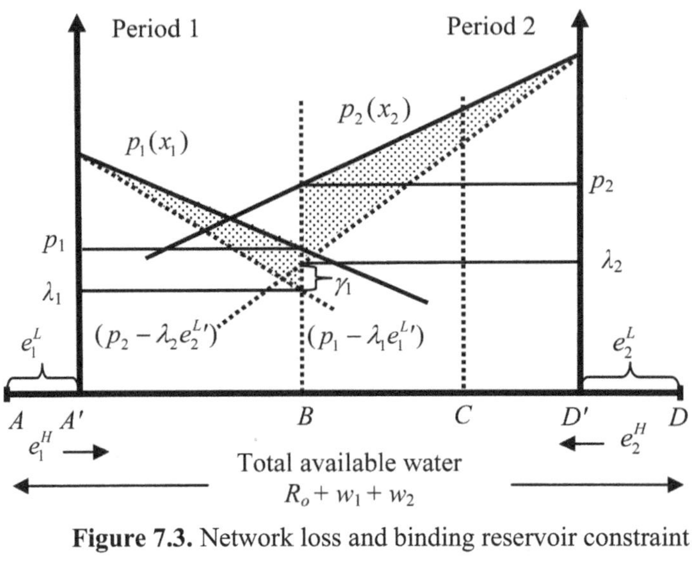

Figure 7.3. Network loss and binding reservoir constraint

ation of the marginal loss in period 1, using the shadow price on the reservoir capacity constraint:

$$\begin{aligned} p_2(x_2) - p_1(x_1) &= \lambda_2 \frac{\partial e_2^L}{\partial x_2} - (\lambda_2 - \gamma_1)\frac{\partial e_1^L}{\partial x_1} \\ &= \lambda_2 \left(\frac{\partial e_2^L}{\partial x_2} - \frac{\partial e_1^L}{\partial x_1}\right) + \gamma_1 \frac{\partial e_1^L}{\partial x_1} \end{aligned} \tag{7.17}$$

Notice that congestion terms are not present in Equation (7.17). It is easy to see from the figure that since both the reservoir constraint and the line constraint restrict the amount of electricity that can be consumed in period 2, both will not in general be binding at the same time. If the line capacity should be binding we do not have to consider the reservoir constraint.

The impact of congestion together with losses is illustrated in Figure 7.4. Consumption is somewhat higher in period 2 than period 1 and constraining the capacity $\bar{x}$ of the line by design of the figure. Congestion in period 2 shifts the curve expressing the difference between the social price and the loss term uniformly downwards with the size of the shadow price on the line capacity, as indicated by the two broken curves below the demand curve for period 2. The optimal value of the water value is found as the intersection of the demand curve for period 1 subtracted the value of the marginal loss and the demand curve for period 2 subtracting both the

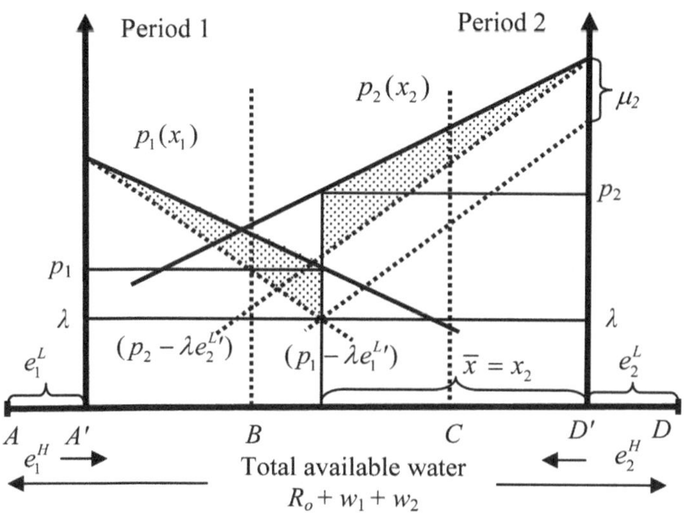

Figure 7.4. Impact of network loss and congestion on social prices

value of the marginal loss and the shadow price on the line-capacity constraint. Since less electricity is consumed in period 2 the loss is now less in this period, as indicated in the figure. The increased consumption in period 1 increases the loss in this period, but the increase must be less than the decrease in period 2, leading to a higher total consumption. But this does not increase the value of the social objective function, on the contrary; we will have a reduction. The reason is that the composition of electricity consumption between the two periods has moved in the wrong direction, as indicated by the increased price difference between the price in the high-demand period 2 and the low demand period 1. The price difference is given by Equation (7.16b) with the shadow price for congestion in period 2 positive and for period 1 zero.

Three nodes and two periods

Let us now extend the model (7.12) to having two generating nodes, but each node with a separate transmission line to the single consumer node. One hydropower producer with a reservoir is assumed to operate at each node. Figure 7.5 provides an illustration. This is an example of the simplest radial network. Furthermore, we assume that one line has greater resistance than the other [in terms of Ohm's R introduced in Equation (7.1) and defined in (7.2)]. This means that a given amount of electricity received at the consumer node generates a greater loss in one line than the other. (The example is due to Wangensteen, 2007.)

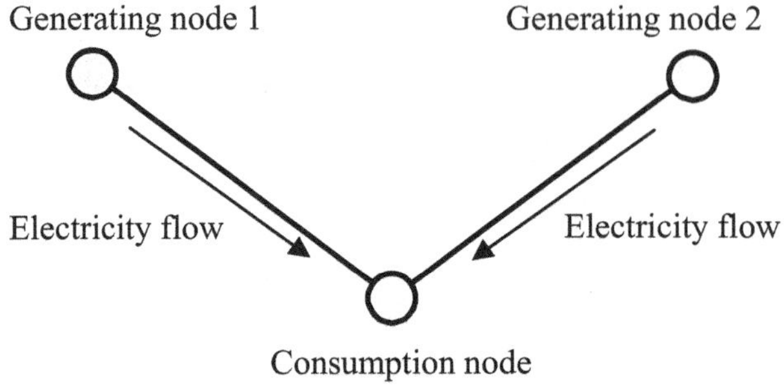

Figure 7.5. Two generation nodes and one consumption node. Radial network

The optimisation problem is the following:

$$\begin{aligned}
&\max \sum_{t=1}^{2} \int_{z=0}^{x_t} p_t(z)dz \\
&\text{subject to} \\
&R_{jt} \le R_{j,t-1} + w_{jt} - e_{jt}^H \\
&R_{jt} \le \overline{R}_j \\
&x_{jt} + e_{jt}^L = e_{jt}^H \\
&x_t = \sum_{j=1}^{2} x_{jt} \\
&e_{jt}^L = e_{jt}^L(x_{jt}) \\
&x_{jt} \le \overline{x}_j \\
&x_t, x_{jt}, e_{jt}^H, e_{jt}^L \ge 0 \\
&w_{jt}, R_{jo}, \overline{R}_j, \overline{x}_j \text{ given}, j = 1,2\ ,\ t = 1,2
\end{aligned} \tag{7.18}$$

Simplifying by substituting for total consumption and loss in each period the Lagrangian is:

$$\begin{aligned}
L = &\sum_{t=1}^{2} \int_{z=0}^{\sum_{j=1}^{2} x_{jt}} p_t(z)dz \\
&-\sum_{t=1}^{2}\sum_{j=1}^{2} \lambda_{jt}(R_{jt} - R_{j,t-1} - w_{jt} + e_{jt}^H) \\
&-\sum_{t=1}^{2}\sum_{j=1}^{2} \gamma_{jt}(R_{jt} - \overline{R}_j) \\
&-\sum_{t=1}^{2}\sum_{j=1}^{2} \tau_{jt}(x_{jt} + e_{jt}^L(x_{jt}) - e_{jt}^H) \\
&-\sum_{t=1}^{2}\sum_{j=1}^{2} \mu_{jt}(x_{jt} - \overline{x}_j)
\end{aligned} \tag{7.19}$$

The necessary first-order conditions are:

$$\frac{\partial L}{\partial x_{jt}} = p_t(x_t) - \tau_{jt} - \tau_{jt}\frac{\partial e_{jt}^L}{\partial x_{jt}} - \mu_{jt} \le 0 \ (= 0 \text{ for } x_{jt} > 0)$$

$$\frac{\partial L}{\partial e_{jt}^{H}} = -\lambda_{jt} + \tau_{jt} \le 0 \; (= 0 \text{ for } e_{jt}^{H} > 0)$$
$$\frac{\partial L}{\partial R_{jt}} = -\lambda_{jt} + \lambda_{j,t+1} - \gamma_{jt} \le 0 \; (= 0 \text{ for } R_{jt} > 0) \tag{7.20}$$
$$\lambda_{jt} \ge 0 (= 0 \text{ for } R_{jt} < R_{j,t-1} + w_{jt} - e_{jt}^{H})$$
$$\gamma_{jt} \ge 0 (= 0 \text{ for} R_{jt} < \overline{R}_{j})$$
$$\mu_{jt} \ge 0 (= 0 \text{ for } x_{jt} < \overline{x}_{j}) \,,\; j = 1,2 \,,\; t = 1,2$$

The shadow prices τ_{jt} on the energy balance for each line are free in sign. We will assume that positive amounts of electricity are consumed in each period. At least one power station must then produce in each period. In fact, both plants will produce in period 2 since this is the terminal period, and we assume no satiation of consumption. For a station producing the first condition in (7.20) holds with equality. This must then also be the case for the second condition. For a power plant not producing in a period the sum of the shadow price on the energy balance, the marginal value of the loss in the transmission to the consumer, and the shadow price for congestion is greater than (or equal to) the social price for the consumer. The second condition tells us that for a plant not producing the water value exceeds (or is equal to) the shadow price on the energy balance.

If both plants produce in both periods we have from the second condition that the water value for a plant for a period must become equal to the shadow price on the energy balance in question for the period. The shadow prices on the energy balances become positive the way we have set up the optimisation problem. The shadow prices are in general both time-specific and plant-specific. The water values are also in general time- and plant-specific. Concerning the latter, we have, from the equation of motion of the shadow prices concerning the reservoirs, that in the case of no threat of overflow (threat of overflow can at most be relevant for period 1), the water value and the energy-balance shadow price for a plant are constant over the periods. However, both types of shadow prices are still different between plants.

Inserting the water values the first-order conditions become:

$$p_t(x_t) = \lambda_j + \lambda_j \frac{\partial e_{jt}^{L}}{\partial x_{jt}} + \mu_{jt} = \lambda_j (1 + \frac{\partial e_{jt}^{L}}{\partial x_{jt}}) + \mu_{jt} ,\; j = 1,2 \,,\; t = 1,2 \tag{7.21a}$$

The common water value over time for a plant is written λ_j. The consumer price is equal to the sum of the water value, the value of the marginal loss, and the shadow price on the line capacity. The loss is evaluated using the water value in question and not the consumer price. This condition is a generalisation of condition (7.15b) in the case of one plant only to two plants. (A further generalisation to N plants is immediate.) The difference between the consumer price and the water value for a plant for each time period is the sum of the loss and the congestion term for the line in question. The water values must be less than the consumer price if either the marginal loss or the shadow prices on congestion are positive.

Since the consumer price is independent of plant the implication of (7.21a) for the relationship between the loss and congestion terms for the plants for the same time period is:

$$\lambda_1(1+\frac{\partial e_{1t}^L}{\partial x_{1t}})+\mu_{1t}=\lambda_2(1+\frac{\partial e_{2t}^L}{\partial x_{2t}})+\mu_{2t},\ t=1,2 \tag{7.21b}$$

The sum of loss-adjusted water values and shadow value of congestion must be equal for the plants for each time period.

The value of the sum of the two terms will in general be different between two periods when demand varies, implying a variation of the social price. The marginal loss term will be higher in the high-demand period than the low-demand period by definition, because marginal loss increases with energy delivered. We then have that both plants will produce more in the high-demand period than they do in the low-demand period.

Occurrence of congestion cannot change this situation in general. We will maintain the assumption in the previous section that line constraints do not lead to locking in of water; $R_{jo}+w_{j1}+w_{j2}<2\bar{x}_j$ ($j=1,2$). It is therefore the case that if congestion occurs on a line, it will be in the high-demand period because production at both plants are higher. But the value of the marginal loss generated by the restricted component of consumption will still be higher than in the low-demand period, and in addition the positive congestion term adds to the cost of the loss. The total effect is that the social price in the high-demand period is higher than the price in the low-demand period.

The difference between the social consumer prices for the two time periods is found by using (7.21a) and is equal to the difference in value of marginal loss and congestion term:

$$p_2(x_2)-p_1(x_1)=\lambda_j(\frac{\partial e_{j2}^L}{\partial x_{j2}}-\frac{\partial e_{j1}^L}{\partial x_{j1}})+(\mu_{j2}-\mu_{j1}),\ j=1,2 \tag{7.22}$$

This relationship is a generalisation of (7.16b). The additional information is that the differences between the sum of the marginal loss term and congestion term for each plant are equal.

Let us first study the total impact on the water use on the periods caused by transmission losses and assume no congestion. If period 1 is the low-demand period and period 2 the high-demand one, at least one plants must produce more in the high-demand period. But then both plants must produce more according to condition (7.22). Furthermore, the higher price in period 2 seen from (7.22) means that more is consumed in period 1 and less in period 2 compared with a situation without transmission losses. We have the same shift of water use from the high-demand period to the low-demand period as in the previous section with one plant.

Concerning the use of water at the plant level let us assume that the marginal loss on line 1 is greater than the marginal loss on line 2 for the same amount of energy delivered at the consumption node. From Ohm's law we have that the second derivative of the loss function is positive [and approximately constant, cf. (7.1)]. The optimality condition (7.21b) demands equality of the loss-adjusted water values for each time period. If we assume that plant 1 has a greater total water inflow than plant 2, then it is reasonable to assume that marginal loss will be greater for plant 1 than plant 2 for both periods, and, consequently the water value for plant 1 will be lower than for plant 2. In fact, the difference in marginal loss values must be of the same sign for both periods, as can be seen from (7.21b).

In order to maintain the equality between loss-adjusted water values, remembering that plant-specific water values are constant over time, plant 1 must have a different relative profile of water use between the periods than plant 2. Because the marginal loss increases more rapidly for plant 1 than for plant 2 the increase in the use of water in period 2 will be relatively less for plant 1 than for plant 2. This means that relatively more water is used in period 1 by plant 1 and less by plant 2. The equality of loss-adjusted water values for each period is obtained by adjusting the relative use of water for each plant between the periods in the fashion described. According to (7.22) the value of the difference between the marginal losses must be the same for each plant. By processing relatively more water in the low-demand period in the plant with the line with the highest Ohm's resistance and relatively less in the high-demand period, and vice versa for the plant with a line with less resistance, the total loss over both periods is reduced compared with a policy of uniform regulation of water use.[3]

[3] These effects are shown numerically in a somewhat simpler model in Wangensteen (2007), assuming equal total inflows for the two plants.

However, the situation described above may be reversed if it is the case that the plant connected through the high-resistance line has less total water to process. If the level effect of resistance is dominated by the volume effect regarding losses, then the relative adjustment for the plants is reversed. It is still the case that relatively more electricity is consumed in the low-demand period and relatively less in the high-demand period compared with a situation without losses.

Considering congestion, the congestion terms may be regarded as constants in (7.21b). If congestion occurs, it will be in the high-demand period. The relative adjustment of production will qualitatively be the same as above, independent of the value of the congestion effect, but the absolute adjustment will be influenced. It seems reasonable that one line only will be congested in period 2. Assuming that the low-resistance line is congested in the high-demand period, but not the high-congestion line ($\mu_{12} = 0$, $\mu_{22} > 0$) leads to a relatively smaller difference in the production between the two plants. The relative difference becomes greater if the high-resistance line is congested, but not the low-resistance line.

A general transmission model

We will now expand the model to encompass N generation nodes, M consumption nodes and S network links. For convenience we label generating nodes the same way as individual generators have been labelled in Chapter 4, but we do not look at individual generators within the same node. We look at aggregated demand for each consumption node. Consumption nodes may coincide with supply nodes, but for simplicity we use separate indices for consumer and producer nodes without specifying if some nodes coincide. Ideally, we would have liked to specify functions that accurately reflect the underlying physical and engineering properties of electricity. However, as mentioned before, this task is complex and will take us too far outside a traditional economic approach. The purpose of the modelling effort here is to maintain a model structure familiar to economists, but still reflecting main features of physical and engineering properties. It will not be shown explicitly how the various links within the transmission network are connected. The network *implicitly* behind the scene is in general exhibiting loop-flows, implying that it is not possible to direct electricity along specific lines. We will capture the physical network implicitly through the generation of losses on each line. These losses are related to generation at all generation nodes and consumption at all consumption nodes.

Keeping our variables in energy units we define the net flow, b_{st}, on a line. We then assume that the generation at each node and the consumption at each node will influence net flow on lines:

$$b_{st} = b_{st}(x_{1t},..,x_{Mt},e_{1t}^{H},..,e_{Nt}^{H})$$
$$t = 1,..,T, s = 1,..,S \tag{7.23}$$

The partial derivatives of this flow relationship may be both positive and negative, and, of course, zero. The equation captures the pervasive electric externalities in a general network; "everything depends on everything."

The losses are then created on each line as a function of the net flow on the line:

$$e_{st}^{L} = e_{st}^{L}(b_{st}), \frac{\partial e_{st}^{L}(b_{st})}{\partial b_{st}} > 0, \frac{\partial^2 e_{st}^{L}(b_{st})}{\partial b_{st}^2} > 0$$
$$t = 1,..,T, s = 1,..,S \tag{7.24}$$

Loss is increasing in line flow. It would be fine if the network could be modelled in a point-to-point way expressing how much electricity is lost in the transport of electricity from a generating node j to a consumer node, i. But loss incurred may be quite impractical to calculate in such a way and also difficult. We therefore stick to a general way of capturing the loss incurred on each line by injections and withdrawals.

In principle electrical equilibrium at all nodes should be modelled: consumption at each node must be equal to the net flow coming in, but we do not identify net flows by nodes. We therefore cannot show the equilibrium at each node. In order to do that we need to specify links into each consumption node and how much electricity that is delivered to the node at the end of each link into the node. The characterisation of power flow over each line is instead implicitly embedded in the energy-balance equation for the total system.

Congestion is also a pervasive phenomenon in a network model. A congested line somewhere may create repercussions throughout the total network. This may be brought out in the simplest possible illustration of a loop-flow possibility using the popular three nodes example; two generation nodes and one consumption node. Adding a link between the two generation nodes in Figure 7.5 we get the ubiquitous triangular model shown in Figure 7.6. The current can either flow directly from a generation node to the consumption node, or flow the other way through the other generation node to the consumption node. The loop-flows are created by the possibility of the flows from the generator nodes to take two different ways to the consumption node. Kirchhoff's laws tell us that the power between any

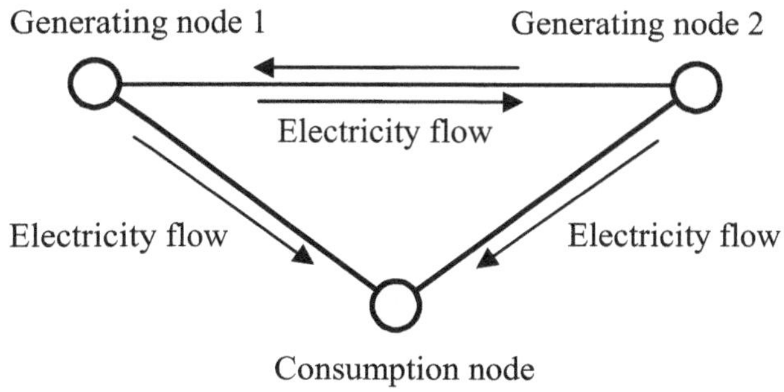

Figure 7.6. Two generation nodes and one consumer node with loop-flows.

two nodes is necessarily distributed across all parallel paths. The distribution on the loops is according to relative resistance on the lines. The size of the flow going directly from a generation node to the consumer node, compared with the flow going the other way through the other generation node, is in proportion to the resistances on these loops. But the really intriguing consequence of the physical laws is that if a flow restriction on a line is reached, then this will determine the maximal flows on the other loop-lines, too. Consider an upper limit on the line between the two generators in Figure 7.6. Then the maximal flow on the direct link between the generation node and the consumption node will be determined by the relative resistances multiplied with the capacity on the link between the generators, even though the capacity on the direct link may be larger.

However, it will take us too far into electrical engineering to try to capture loop-flow externalities. We will model line capacities as given, and then let all injections and all withdrawals influence the flow on lines:

$$\begin{aligned} &b_{st} = b_{st}(x_{1t},..,x_{Mt},e_{1t}^H,..,e_{Nt}^H) \le \bar{b}_s \\ &t = 1,..,T, s = 1,..,S \end{aligned} \qquad (7.25)$$

This formulation cannot capture the loop-flow congestion externalities illustrated by Figure 7.6, because a binding constraint on a line there may reduce feasible upper levels on other lines below their physical limits.

Another source of transmission constraint in addition to the thermal aspect is the voltage. Reactive power occurs on an alternating current network creating restrictions and also voltage stability problems. A complete analysis of the network requires modelling both real and reactive power. However, we will not attempt to include such issues here.

The general social planning problem with a transmission network can then be formulated:

$$\begin{aligned}
&\max \sum_{t=1}^{T}\sum_{i=1}^{M}\int_{z=0}^{x_{it}} p_{it}(z)dz \\
&\text{subject to} \\
&R_{jt} \le R_{j,t-1} + w_{jt} - e_{jt}^{H} \\
&R_{jt} \le \overline{R}_{j} \\
&\sum_{i=1}^{M} x_{it} + \sum_{s=1}^{S} e_{st}^{L} = \sum_{j=1}^{N} e_{jt}^{H} \\
&e_{st}^{L} = e_{st}^{L}(b_{st}) \\
&b_{st} = b_{st}(x_{1t},..,x_{Mt},e_{1t}^{H},..,e_{Nt}^{H}) \\
&b_{st} \le \overline{b}_{s} \\
&R_{jt}, x_{it}, e_{jt}^{H}, e_{st}^{L}, b_{st} \ge 0 \\
&T, w_{jt}, R_{jo}, \overline{R}_{j}, \overline{b}_{s} \text{ given} \\
&t = 1,..,T,\ j = 1,..,N,\ i = 1,..,M,\ s = 1,..,S
\end{aligned} \tag{7.26}$$

Inserting for loss the Lagrangian is:

$$\begin{aligned}
L = &\sum_{t=1}^{T}\sum_{i=1}^{M}\int_{z=0}^{x_{it}} p_{it}(z)dz \\
&-\sum_{t=1}^{T}\sum_{j=1}^{N}\lambda_{jt}(R_{jt} - R_{j,t-1} - w_{jt} + e_{jt}^{H}) \\
&-\sum_{t=1}^{T}\sum_{j=1}^{N}\gamma_{jt}(R_{jt} - \overline{R}_{j}) \\
&-\sum_{t=1}^{T}\tau_{t}\left(\sum_{i=1}^{M} x_{it} + \sum_{s=1}^{S} e_{st}^{L}(b_{st}(x_{1t},..,x_{Mt},e_{1t}^{H},..,e_{Nt}^{H})) - \sum_{j=1}^{N} e_{jt}^{H}\right) \\
&-\sum_{t=1}^{T}\sum_{s=1}^{S}\mu_{st}(b_{st}(x_{1t},..,x_{Mt},e_{1t}^{H},..,e_{Nt}^{H}) - \overline{b}_{s})
\end{aligned} \tag{7.27}$$

The necessary first-order conditions are:

$$\frac{\partial L}{\partial x_{it}} = p_{it}(x_{it}) - \tau_{t}\left(1 + \sum_{s=1}^{S}\frac{\partial e_{st}^{L}}{\partial b_{st}}\frac{\partial b_{st}}{\partial x_{it}}\right) - \sum_{s=1}^{S}\mu_{st}\frac{\partial b_{st}}{\partial x_{it}} \le 0 \ (= 0 \text{ for } x_{it} > 0)$$

$$\frac{\partial L}{\partial e^H_{jt}} = -\lambda_{jt} + \tau_t(1 - \sum_{s=1}^{S} \frac{\partial e^L_{st}}{\partial b_{st}} \frac{\partial b_{st}}{\partial e^H_{jt}}) - \sum_{s=1}^{S} \mu_{st} \frac{\partial b_{st}}{\partial e^H_{jt}} \le 0 \; (= 0 \text{ for } e^H_{jt} > 0)$$

$$\frac{\partial L}{\partial R_{jt}} = -\lambda_{jt} + \lambda_{j,t+1} - \gamma_{jt} \le 0 (= 0 \text{ for } R_{jt} > 0) \tag{7.28}$$

$$\lambda_{jt} \ge 0 (= 0 \text{ for } R_{jt} < R_{j,t-1} + w_{jt} - e^H_{jt})$$

$$\gamma_{jt} \ge 0 (= 0 \text{ for} R_{jt} < \bar{R}_j)$$

$$\mu_{st} \ge 0 (= 0 \text{ for } b_{st} < \bar{b}_s)$$

$$i = 1,.., M,\ t = 1,.., T, j = 1,.., N,\ s = 1,.., S$$

The shadow price τ_t on the energy balance is free in sign. Looking at the number of endogenous variables and equations, the endogenous variables may in principle be determined, but due to the somewhat unclear properties of the line-flow function the sufficiency conditions may be violated, indicating that there may be problems with attaining a unique optimum.

We will assume that there is positive consumption at each consumer node, implying that the first condition holds with equality. The social consumer price at node i can then be expressed as

$$p_{it}(x_{it}) = \tau_t + \tau_t \sum_{s=1}^{S} \frac{\partial e^L_{st}}{\partial b_{st}} \frac{\partial b_{st}}{\partial x_{it}} + \sum_{s=1}^{S} \mu_{st} \frac{\partial b_{st}}{\partial x_{it}} \;,\; i = 1,.., M,\ t = 1,.., T \tag{7.29}$$

The first term on the right-hand side is the shadow price on the energy balance. This is thc opportunity cost of the unit increase in consumption at node i. The second term on the right-hand side is expressing the marginal losses on all the S lines created due to the marginal increase in consumption at node i evaluated using the shadow price on the energy balance. Given an increase of the flow on line s the loss is increasing, but flows on lines may go up as well as down when consumption at node i increases marginally. Therefore the total expression for loss may be positive as well as negative. This is also the case for the expression for congestion. However, the congestion term cannot be negative for all consumer nodes if one of the constraints is binding. We would expect as a normal result that the majority of the expressions are positive. One must be careful not to confuse a characterisation of the optimal solution with some line constraints being binding. A consumer node located in, e.g., a locked-in export region may have a negative congestion term, but the shadow price on a congested link out of the region may still remain positive. The consumption in the export region will increase compared with an unconstrained case due to a lower consumer price, and the congestion is thereby not relieved to the extent that the shadow price on the link becomes zero.

The shadow price on the energy balance is free in sign since the energy balance is an equality constraint. We should find the shadow price positive the way we have set up the problem. If the loss decreases more than the unit increase in consumption at a node, then it might seem possible that the social price becomes negative if the loss term outweighs the sum of the shadow price on the energy balance and the congestion term. The consumers at the node would then be paid to use more electricity. However, since the shadow price on the energy constraint is common for all consumer and generating nodes it seems rather impossible that all the nodes are characterised by having negative losses. We will therefore adopt the assumption that the shadow price on the energy-balance constraint is positive. It is still possible for a consumption node to have a negative social price.

In the case of no losses being created and no binding line capacity constraints, the social consumer price equals the shadow price on the energy balance constraint as in the corresponding models of the previous chapters.

Assuming that there is positive generation at node j the water value becomes

$$\lambda_{jt} = \tau_t - \tau_t \sum_{s=1}^{S} \frac{\partial e_{st}^L}{\partial b_{st}} \frac{\partial b_{st}}{\partial e_{jt}^H} - \sum_{s=1}^{S} \mu_{st} \frac{\partial b_{st}}{\partial e_{jt}^H}, j = 1,..,N, t = 1,..,T \tag{7.30}$$

The water value equals the shadow price on the energy balance subtracted system losses created at the margin due to the unit injection, valued at the shadow price of the energy balance, and the shadow-valued congestion costs. If both the loss and the congestion terms are positive the water value becomes smaller than the shadow price on the energy balance. The water value must be non-negative. We see from the second condition in (7.28) that if the water value remains larger than the difference between the shadow price on the energy balance and the sum of loss and congestion terms for all feasible values of production, then production is set to zero in this period. As was the case for a consumption node the loss term may now also be negative, making the stored water at the generation node more valuable. This may be the case of a generation node being the closest to a large consumer node. The congestion term may also be negative contributing to an increase in the water value. This may be the case for a generating unit within an import-restricted region.

If losses and congestion are zero the water value becomes equal to the shadow price on the energy balance, implying as in the models of Chapter 4 that the water values are all the same and equal to the common water value of active generators.

The role of a comprehensive loss and congestion social pricing can be seen by inspecting the pair-wise differences between prices at consumer nodes, prices at generating nodes, and prices between a consumer and a generating node. The difference between social consumer prices at two nodes i and u is found using Equation (7.29):

$$p_{it}(x_{it}) - p_{ut}(x_{ut}) = \tau_t(\sum_{s=1}^{S} \frac{\partial e_{st}^L}{\partial b_{st}} \frac{\partial b_{st}}{\partial x_{it}} - \sum_{s=1}^{S} \frac{\partial e_{st}^L}{\partial b_{st}} \frac{\partial b_{st}}{\partial x_{ut}}) + \sum_{s=1}^{S} \mu_{st} \frac{\partial b_{st}}{\partial x_{it}} - \sum_{s=1}^{S} \mu_{st} \frac{\partial b_{st}}{\partial x_{ut}} = \sum_{s=1}^{S} (\tau_t \frac{\partial e_{st}^L}{\partial b_{st}} + \mu_{st})(\frac{\partial b_{st}}{\partial x_{it}} - \frac{\partial b_{st}}{\partial x_{ut}}) \tag{7.31}$$
$$i,u = 1,..,M,\ t = 1,..,T$$

A higher loss and a higher congestion at one node compared with another contributes to the former node having the highest social consumer price. Consumers located at a node generating higher losses and congestion at the margin should get incentives to scale back consumption. The general case is that all social prices are different. The social prices between pairs of consumption nodes will only become equal if the loss and congestion effects at the margin are identical.

In an analogous way the difference in water values between a pair of generating nodes j and v can be found using (7.30):

$$\lambda_{jt} - \lambda_{vt} = \tau_t(\sum_{s=1}^{S} \frac{\partial e_{st}^L}{\partial b_{st}} \frac{\partial b_{st}}{\partial e_{vt}^H} - \sum_{s=1}^{S} \frac{\partial e_{st}^L}{\partial b_{st}} \frac{\partial b_{st}}{\partial e_{jt}^H}) + \sum_{s=1}^{S} \mu_{st} \frac{\partial b_{st}}{\partial e_{vt}^H} - \sum_{s=1}^{S} \mu_{st} \frac{\partial b_{st}}{\partial e_{jt}^H}) = \sum_{s=1}^{S} (\tau_t \frac{\partial e_{st}^L}{\partial b_{st}} + \mu_{st})(\frac{\partial b_{st}}{\partial e_{vt}^H} - \frac{\partial b_{st}}{\partial e_{jt}^H}) \tag{7.32}$$
$$j,v = 1,..,N,\ t = 1,..,T$$

The generation node with highest sum of total loss and congestion at the margin will have the lowest water value. Generation at such nodes become cheaper in terms of opportunity cost of water.

The difference between the nodal social price at a consumer node i and the water value of a generating node j is found by combining (7.29) and (7.30):

$$p_{it}(x_{it}) - \lambda_{jt} = \tau_t(\sum_{s=1}^{S} \frac{\partial e_{st}^L}{\partial b_{st}} \frac{\partial b_{st}}{\partial x_{it}} + \sum_{s=1}^{S} \frac{\partial e_{st}^L}{\partial b_{st}} \frac{\partial b_{st}}{\partial e_{jt}^H}) + \sum_{s=1}^{S} \mu_{st} \frac{\partial b_{st}}{\partial x_{it}} + \sum_{s=1}^{S} \mu_{st} \frac{\partial b_{st}}{\partial e_{jt}^H} = \sum_{s=1}^{S} (\tau_t \frac{\partial e_{st}^L}{\partial b_{st}} + \mu_{st})(\frac{\partial b_{st}}{\partial x_{it}} + \frac{\partial b_{st}}{\partial e_{jt}^H}) \tag{7.33}$$

$i = 1,..,M, j = 1,..,N, t = 1,..,T$

The difference is the sum of the two loss terms evaluated using the shadow price of the energy balance and the two congestion terms, evaluated by the shadow prices of the line capacity constraints. When the loss and congestion terms are positive the social price is greater than the water value for all relevant pairs of consumer and generating nodes.

Separation into zones

Congestion may lead to separation of a system covered by a grid into zones that become independent as to price formation. An example is provided in Figure 7.7. The generating nodes are indicated with small circles and the consumption nodes with large circles. The meshed grid pattern just indicates that there are several ways for the electricity to flow from production nodes to consumption nodes, i.e., loop flows may occur. Size of generation and demand, or capacities of links are not indicated in the figure. The network falls in two parts; the southern and the northern parts,

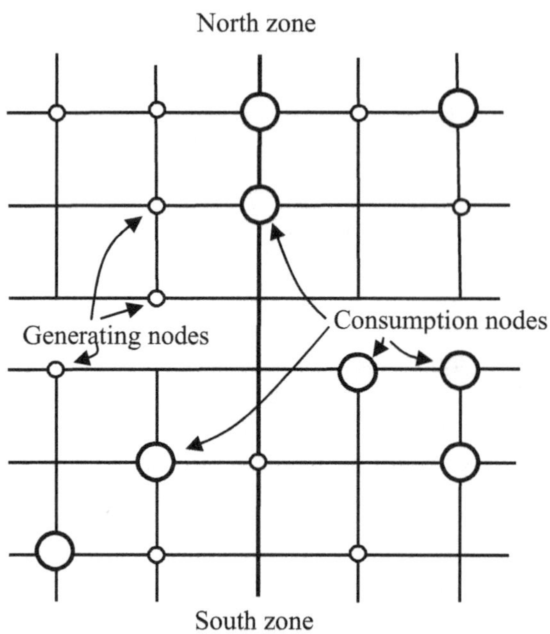

Figure 7.7. A general transmission network. Generating nodes are represented by small circles, consumption nodes are represented by large circles.

and there is only a single link between these two parts. This is the case between the North and the South Island of New Zealand, and almost the case between North and South Norway. This connecting link may be congested for certain configurations of supply and demand in the total network. Usually there are critical links with restricted capacity that cause congestion. But these links may change with demand and supply configurations. Notice that with loop-flows we may have congestion occurring without resulting in separate zones. Such separation was assumed in Chapter 6 when looking at two countries, and separate prices resulted when the link between the countries was congested with the importing country having the highest prices.

When a grid is separated by congestion the determination of social prices, water values, and other shadow prices will take place insulated from events in the other parts. Equations (7.29) - (7.33) will all be zone-specific. The externalities exhibited will only contain elements related to generation and consumption nodes within the zone and to links within the zones, forming subsets of the general sets of the consumption and generating nodes and lines.

Network impact on utilisation of hydropower

The nodal price structure due to loss and congestion externalities as revealed above is general and valid for various types of generators. The water value applying to a generator node represents marginal generation cost for hydropower. It is shown by Equations (7.30) and (7.32) that water values are in general different. What is special for hydropower is the dynamics of the shadow prices of water and reservoir limits as revealed in the third condition in (7.28). As long as reservoir levels stay in between empty and full the water value remains constant. The three elements shadow price on the energy balance, value of total marginal losses, and congestion may change from period to period, but the water value remains the same.

The simple examples of two and three nodes revealed that the pattern of use of reservoirs is influenced by transmission. Less water will be used in high-demand periods due to the increased losses incurred. Differential losses on lines will also influence the relative use of hydropower plants connected to consumer nodes with different resistances, e.g., due to different geographical distances. In our example reservoirs connected through lines with less resistance will be used relatively more extensively in high-demand periods than low-demand periods, and vice versa for reservoirs connected through lines that have relative higher resistance.

Comparing the nature of the optimal solution with transmission to the ones without in Chapter 4 we no longer obtain uniform water values in the system, but generating-node specific water values. Furthermore, the prices at consumer nodes become in general different. The differences in water values and in consumer prices all stem from the way losses are incurred in the system and effects of congestion. Congestion is even more important than we have modelled due to loop-flow effects. This is the background for proposals of *spot-pricing* (Bohn et al., 1984; Schweppe et al., 1988).[4] An important implication for the social planning solution is that Hveding's conjecture cannot be invoked to aggregate the system. The spatial distribution of dispatch of generators within a period must take losses into consideration, created simultaneously by the spatial distribution of demand. The utilisation profile of reservoirs over time will be influenced by spatial variation in losses. When consideration of overflow necessitates a specific manoeuvring of a reservoir, the creation of loss connected to the utilisation profile will also enter the picture.

However, it should be evident from our analysis and the physical electrical realities that, even for a social planner, it would be quite an involved operation in practice in real time to mirror the physical system completely by fully implementing the spatial structure of social prices at consumer nodes and individual water values at generating nodes that takes incurred losses and congestion fully into consideration. The transaction costs in the form of gathering information, processing it, and sending instructions to generators may involve costs that are higher than the social benefit of spatial pricing. The way the electricity flow from one generating node is distributed on consumption nodes varies continuously over time and with the changes in the configurations of consumption and generation, thus creating "electrical externalities" of losses and congestion involving loop-flow effects in the network system. It may be impractical, or too costly, to internalise the full extent of externalities.

Our analysis can provide an understanding of assumptions that have to be made in order for equal water values to be faced by producers, and equal social prices by consumers. The general condition is uniformity of marginal loss effects and congestion impact over generation and consumption nodes. Then prices are equal and water values are equal, and there is a constant mark-up factor between water values and consumer prices. But this approximation may be too crude to follow in practice. Losses and transmission constraints in looped networks are likely to generate significant interaction effects across different parts of the system and lead to a

---

[4] According to Bohn et al. (1984) spot-pricing was first proposed in Vickrey (1971) as "responsive pricing."

social price structure of different nodal prices and water values. The analysis above may shed some light on design of spatial pricing and benefits to be reaped (Green, 2007).

Chapter 8. Market Power

The deregulation of the electricity power production system in many countries since the early 1990s has stimulated interest in the possibilities of producers behaving strategically. The classical implication of use of market power that production is reduced compared with perfect competition also holds for electricity markets being supplied by conventional thermal power. Typical base-load plants like nuclear power plants do not have the same physical opportunities because of long and expensive start-up and close-down times. Systems with a significant contribution from hydro power with storage of water have not been studied so much. However, hydropower plays a significant part in many countries. As pointed out in Chapter 1 about 20% of the world's electricity is produced by hydro power, and one third of countries in the world depend on hydropower for more than 50% of their electricity generation (www.hydropower.org). Hydropower with water storage has features that set it apart from other generating technologies concerning possibilities of exercising market power. The almost costless instantaneous change in hydro generation within the power capacities makes it perfect for strategic actions in competition with thermal generators, with both costs and time lags involved in changing production levels of the latter. In countries with day-ahead spot markets hydro producers interact daily and they all know that operating output-depending costs are zero, the opportunity cost is represented by future expected market prices, and they may hold quite similar expectations. This may facilitate collusion. In the case of hydropower, production can be reduced only by using less water. This may lead to spillage of water when reservoirs are limited and inflows positive. Spilling water has the same logic as burning coffee beans to support the coffee price of a cartel, but it is also easily as observable and may be met with regulatory action. Spilling water is obviously not part of a social solution (if technically avoidable), as demonstrated in earlier chapters. One reason for concern about potential market power abuse of hydro producers is that it may be used without any spilling of water and not so easy to detect by regulators, because market power is typically exercised by a *reallocation* of release of water between periods compared with what would be the socially desired release pattern. Measuring existence of market power by comparing price and marginal costs does not work for hydropower because variable cost is virtually zero. The relevant variable cost is the opportunity cost of water, but this is an expected variable and not directly observable.

Although there is some recent literature covering market power by hydro producers, the topic deserves a closer scrutiny and systematic review. Use of market power by hydro producers is covered in Ambec and Doucet (2003) and Crampes and Moreaux (2001) using very simplifying assumptions. Two-periods are considered in both models and the standard result of a monopoly following the strategy of equalising marginal revenues of the periods, resulting in a reallocation of water from periods with relative inelastic demand to periods with relatively more elastic demand, is established. A constraint on the transferability of water from one period to the next is not considered. Borenstein et al. (2002) investigated the possible use of market power by hydro producers when thermal capacities are also present at the backdrop of the California crisis. The formal model is the same as the model in Bushnell (2003) dealing with strategic scheduling of the hydro producer with different assumptions about the behaviour of the thermal producers. When a monopolist controls thermal capacities, the equalisation of the marginal-revenues rule over the periods is confirmed.

Monopoly

In order to expose the strategies of a monopolist we start with the simplest possible case and then increase the complexities later. As a starting point we assume that all hydro producers are part of a monopoly and simplify further by considering the monopolist as a single production unit (i.e., the coordination problems shown in Chapter 4 and summed up as Hveding's conjecture are solved by the monopolist). We assume that the monopolist knows the period demand functions just like the social planner. The optimisation problem of the monopolist in the basic case of a single water availability constraint is:

$$
\begin{aligned}
&\max \sum_{t=1}^{T} p_t(e_t^H) \cdot e_t^H \\
&\text{subject to} \\
&\sum_{t=1}^{T} e_t^H \leq W \\
&T, W \text{ given}
\end{aligned}
\tag{8.1}
$$

The function $p_t(e_t^H)$ is the demand function on price form for period t with standard properties.

The Lagrangian for problem (8.1) is:

$$\sum_{t=1}^{T} p_t(e_t^H) \cdot e_t^H - \lambda(\sum_{t=1}^{T} e_t^H - W) \tag{8.2}$$

The necessary first-order conditions are:

$$\begin{aligned} &\frac{\partial L}{\partial e_t^H} = p_t'(e_t^H)e_t^H + p_t(e_t^H) - \lambda \le 0 \ (= 0 \text{ for } e_t^H > 0),\ t = 1,..,T \\ &\lambda \ge 0 \ (= 0 \text{ for } \sum_{t=1}^{T} e_t^H < W) \end{aligned} \tag{8.3}$$

Assuming that the monopolist will produce electricity in all periods the conditions may be written:

$$p_t(e_t^H)(1 + \breve{\eta}_t) = p_{t'}(e_{t'}^H)(1 + \breve{\eta}_{t'}) = \lambda,\ t,t' = 1,..,T \tag{8.4}$$

In the expression for the marginal revenue of increasing production we have introduced the *demand flexibility*, $\breve{\eta}_t = p_t' e_t^H / p_t$ (the inverse of the demand elasticity), which is negative. The condition is that the marginal revenues, expressed as *flexibility-corrected prices*, should be equal for all the periods and equal to the shadow price on stored water. As in the text-book monopoly case the absolute value of the demand flexibilities (demand elasticities) must be less (greater) than, or equal to, one for a unique solution to exist. The short-run demand may in general be on the inelastic side, so the condition on the price elasticities is not necessarily so innocent. Prices may become quite high in order for the monopolist to be able to push demand to the elastic part of the demand function, and in the case of inelastic demand with vertical demand curve the monopoly solution characterised by (8.3) does not exist. Equality of marginal revenues between periods implies that the period with the relatively most elastic demand at the optimal quantity of electricity, i.e., the smallest absolute value of the demand flexibility $\breve{\eta}_t$, obtains the smallest market price. From 8.4 we have:

$$\begin{aligned} &p_t(e_t^H) = p_{t'}(e_{t'}^H)\frac{1 + \breve{\eta}_{t'}(e_{t'}^H)}{1 + \breve{\eta}_t(e_t^H)} \Rightarrow \\ &p_t(e_t^H) < p_{t'}(e_{t'}^H) \text{ if } \left|\breve{\eta}_t(e_t^H)\right| < \left|\breve{\eta}_{t'}(e_{t'}^H)\right|,\ t,t' = 1,..,T, t \ne t' \end{aligned} \tag{8.5}$$

The benchmark social planning case uses consumer and producer surplus, $\sum_{t=1}^{T}\int_{z=0}^{e_t^H} p_t(z)dz$, as objective function while the monopolist only considers producer surplus, $\sum_{t=1}^{T} p_t(e_t^H)e_t^H$. The difference between the monopoly solution (8.3) or (8.4) and the social solution is that the flexibility-corrected price is substituted for the price. Compared with the solution in the social planning case the monopolist can only obtain higher profit than by using the common social price [marginal willingness to pay in the condition (2.6) in Chapter 2] if the demand functions differ over periods. If the demand functions are identical for the periods it follows from (8.3) that the flexibility-corrected prices become equal, and therefore the prices will be equal and equal to the common price in the social solution, provided that there is no spilling. With spilling the monopoly prices will be equal, but higher than the social prices for identical demand functions. However, the shadow value on the water resource becomes less than this price, reflecting that a monopolist considers the marginal revenue as the opportunity cost of using water. This difference may have implications in a dynamic setting of investment in new capacity. A monopoly will tend to expand less facing, e.g., positive shifts in demand.

If water is left unused we have from (8.3) that the shadow price of water is zero. Since the shadow price of water is a scalar this implies that the

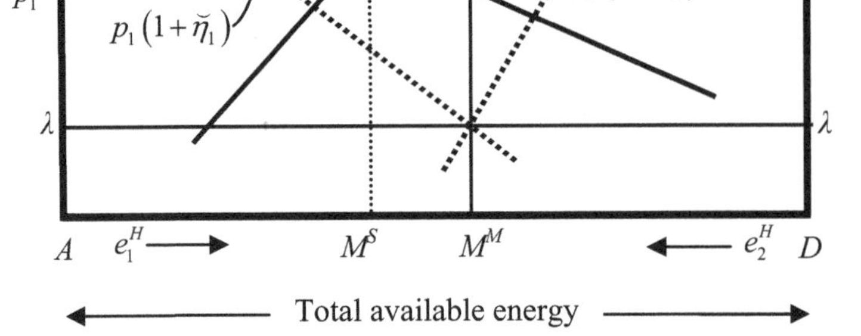

Figure 8.1. The basic monopoly case.
Social solution shown by thin dotted lines.

flexibility-corrected prices must be equal to zero for all periods and hence the price flexibilities equal to 1.

An illustration in the case of two periods, with the same linear demand curves as in Figure 2.1 and the same total water resource, is provided in Figure 8.1. The broken lines are the marginal revenue curves. The length *AD* of the floor of the bathtub indicates the available water. We have that in the illustration the marginal revenue curves intersect at a positive value, i.e., it will not be optimal for the monopolist to leave any water unused. This value is the shadow value on water. But this result depends on the form of the demand functions. If we have unused water as an optimal solution, then the shadow water value is zero. Going vertically up to the demand curves from the intersection point of the marginal revenue curves gives us the monopoly prices for the two periods.

In Figure 8.1 the social solution is indicated by the thin dotted horizontal line $p_1^S p_2^S$ and the corresponding water allocation by the point M^S. The shadow value of water is smaller in the monopoly case than in the social optimal case. If all water is to be used we must have in general that at least one monopoly price is lower than the social price. (Notice that this is not sufficient for all water to be used.) In this case, for the quantity corresponding to the lowest monopoly price the marginal revenue must be lower than the social price for the period in question and consequently the common shadow value on water in the monopoly case must in general be smaller than the shadow value in the social planning case. If water remains unused we have that the shadow value of water is zero, according to the complementary slackness condition in (8.3).

An important general result is that in the case of monopoly the market prices become *different* for the periods, in contrast to the constant price in the social optimal solution indicated by the dotted horizontal line $p_1^S\, p_2^S$. For the period with the most inelastic demand, period 2, the price becomes higher than the social optimal price, and for the most elastic period, period 1, the price becomes smaller, in accordance with (8.5). Thus we have a general *shifting* in the utilisation of water from periods with relative inelastic demand to periods with relative elastic demand. The water allocation in Figure 8.1 moves from point M^S in the social case to M^M in the monopoly case. Although the total electricity supply over the two periods is the same as in the social case, the monopolist increases his profit by selling more in the most elastic period, and then partially reducing his revenue indicated by the marked area $(p_1^S - p_1^M)AM^M$ on the sales in period 1, but recouping more than this in increased revenue in period 2, indicated by the marked area $(p_2^M - p_2^S)M^M D$.

The monopolist will leave water unused if it is optimal to set marginal

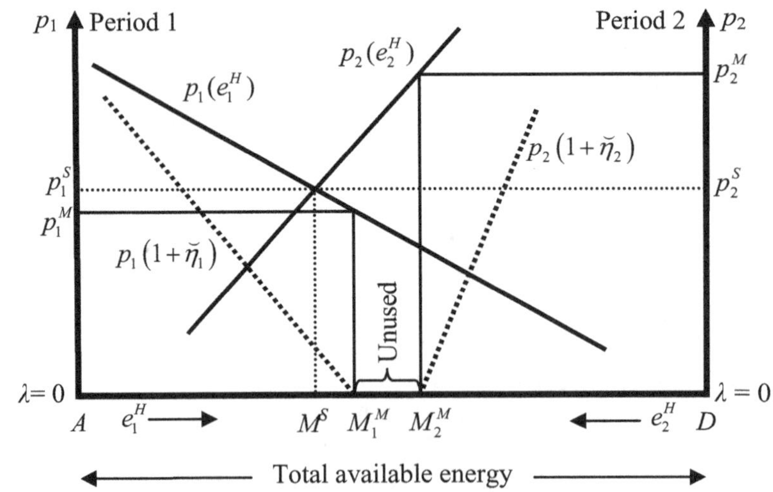

Figure 8.2. Unused water in the monopoly case.
Social solution shown by thin dotted lines.

revenues equal to zero. Note that since we have only one shadow price on the water resource, if marginal revenue is to be zero in one period the marginal revenues have to be zero in all the other periods, too, when water is used in all periods. By changing the slope of the demand curves in Figure 8.1 slightly this case is illustrated in Figure 8.2. The marginal revenue curves do not intersect within the bathtub, and becomes zero at M_1 and M_2 respectively for the two periods. Period 1 has the relatively most elastic demand and more electricity is sold than in the social solution, reducing the monopoly price below the social price, as indicated by the position of the horizontal dotted line for the social case. The available water is not fully utilised; the amount $M_1^M M_2^M$ is left unprocessed. The monopoly price is far above the social price in period 2.

Since unused water is easy to observe it may be of interest to see what the monopoly solution will be if a condition of full use of the available water is made. Technically this means that the water resource constraint is made into an equality constraint so the sign on the shadow price λ in (8.2) is not restricted anymore and the last condition in (8.3) is dropped. Marginal revenues should still be equal and equal to the water shadow price. Using the same demand functions and total water availability as in Figure 8.2 the solution with the water constraint as an equality constraint means that the marginal revenues become negative, and more water is used in

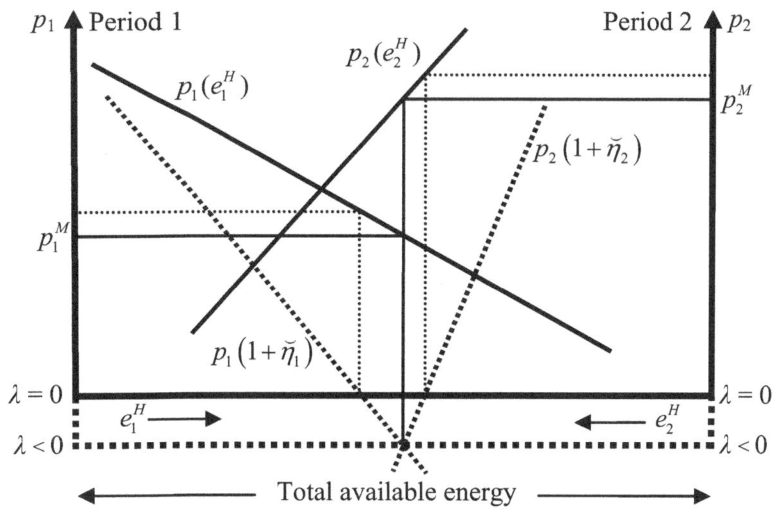

Figure 8.3. Monopoly with full resource-use constraint. Solution without constraint shown by dotted thin lines

both periods, resulting in lower prices in both periods and still unequal prices, as shown in Figure 8.3.

Monopoly and trade

A hydro region with a regional monopoly may engage in electricity trade with neighbouring regions. Let us call a region for a country for ease. We will look at a situation where the monopolist controls both import and export, but takes the import/export prices as given. Unlimited trade will be assumed. Although this is unrealistic it will serve as a benchmark for introducing restrictions on the interconnector capacity later. Extending model (8.4) we have the monopoly profit maximisation problem adding export revenues or subtracting import outlays from the home profit function:

$$\max \sum_{t=1}^{T} p_t(x_t)x_t + p_t^{XI} e_t^{XI}$$

subject to

$$x_t = e_t^H - e_t^{XI}$$
$$\sum_{t=1}^{T} e_t^H \le W \qquad (8.6)$$
$$T, W, p_t^{XI} \text{ given, } e_t^{XI} \text{ free, } t = 1,..,T$$

Here p_t^{XI} is the export/import price (prices are equal and transmission cost is disregarded) and e_t^{XI} is export if positive and import if negative. The first restriction in (8.6) is the energy balance; the consumption x_t at home may be supplied by locally produced hydro or by imports. Inserting the energy balance that holds as an equality constraint yields the Lagrangian:

$$L = \sum_{t=1}^{T} p_t(e_t^H - e_t^{XI}) \cdot (e_t^H - e_t^{XI}) + p_t^{XI} e_t^{XI}$$
$$-\lambda(\sum_{t=1}^{T} e_t^H - W) \qquad (8.7)$$

The necessary first-order conditions are:

$$\frac{\partial L}{\partial e_t^H} = p_t'(e_t^H - e_t^{XI}) \cdot (e_t^H - e_t^{XI}) + p_t(e_t^H - e_t^{XI}) - \lambda \le 0$$
$$(= 0 \text{ for } e_t^H > 0)$$
$$\frac{\partial L}{\partial e_t^{XI}} = -p_t'(e_t^H - e_t^{XI}) \cdot (e_t^H - e_t^{XI}) - p_t(e_t^H - e_t^{XI}) + p_t^{XI} = 0 \qquad (8.8)$$
$$\lambda \ge 0 \; (= 0 \text{ for} \sum_{t=1}^{T} e_t^H < W)$$

We assume that the amount of electricity consumed locally is positive in all periods (i.e., $x_t > 0$) and that the export/import prices are all different. The second condition in (8.8) holds with equality because the export/import variable is not constrained in sign. Because there is an export opportunity to positive price water will not be wasted by the monopolist and the shadow price on water will be positive. If hydro is used in an import period then the first condition in (8.8) holds with equality, implying that the flexibility-corrected home market price, $p_t(1+\breve{\eta}_t)$, is equal to the shadow price on water. The second condition tells us that the flexibility-corrected price is always equal to the import price. But since the export/import prices are different the shadow price on water can be determined only by *one* flexibility-corrected price. We know that in an export period we must also use hydro at home because of the assumption of positive consumption at home of electricity in all periods. Therefore in an export period the flexibility-corrected price is also equal to the shadow price on water. Because of lack of any restriction on trade it is the highest export price period that will become the *only* export period, and in all other periods there will be imports and no use of hydro at home (i.e., no electricity will be produced using water at home). This means that in import periods the flexibility-corrected price is *less* than the shadow price on water.

An illustration is provided in Figure 8.4. Because the import price by construction is lowest in period 1 this period will be the import period. The amount of import is determined by the intersection of the marginal revenue curve and the import price line. The home market price will be higher than the import price in the standard way of a monopoly. Import may be regarded as an alternative way to using hydro to "produce" electricity (marginal revenue is set equal to the marginal production cost; the import price). In the export period the use at home of hydro is determined by the intersection of the marginal revenue curve and the export price line. Export is residually determined as the rest of the available water. The shadow price of water is equal to the export price. Comparing the monopoly solution with the socially optimal solution, the latter is indicated by the vertical dotted lines from the intersection of period 1 demand curve with the

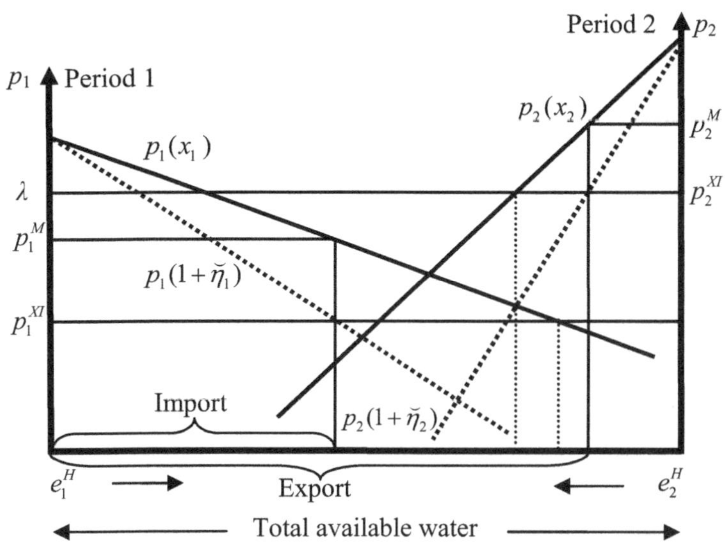

Figure 8.4. Monopoly and trade without restrictions. Social solution shown by vertical dotted lines.

import price for this period, and the intersection of period 2 demand curve with the export price for this period. The import and export periods will be the same. The shadow price on water will be the same in the two solutions, but import will be considerably reduced in the monopoly case, resulting in a higher home price than the import price. In the export period the monopoly will export more water and restrict correspondingly the use of water for electricity production at home, resulting in a home price higher than the export price. The monopolist is playing price discrimination between two markets.

Constraining the amount traded due to limited interconnector capacity makes for a more realistic situation. The monopoly profit maximisation problem in the case of restrictions on trade is:

$$\begin{aligned}
&\max \sum_{t=1}^{T} p_t(x_t)x_t + p_t^{XI} e_t^{XI} \\
&\text{subject to} \\
&x_t = e_t^H - e_t^{XI} \\
&\sum_{t=1}^{T} e_t^H \le W \\
&-\bar{e}^{XI} \le e_t^{XI} \le \bar{e}^{XI},\ e_t^{XI} \text{ unrestricted in sign} \\
&T, W, p_t^{XI}, \bar{e}^{XI} \text{given},\ t = 1,..,T
\end{aligned} \tag{8.9}$$

The restriction on trade can be expressed by one restriction on export and another on imports, remembering that import is negative and export positive. Inserting the energy balance that holds as an equality constraint yields the Lagrangian:

$$\begin{aligned}
L = &\sum_{t=1}^{T} p_t(e_t^H - e_t^{XI}) \cdot (e_t^H - e_t^{XI}) + p_t^{XI} e_t^{XI} \\
&-\lambda\left(\sum_{t=1}^{T} e_t^H - W\right) \\
&-\sum_{t=1}^{T} \alpha_t (e_t^{XI} - \bar{e}^{XI}) \\
&-\sum_{t=1}^{T} \beta_t (-e_t^{XI} - \bar{e}^{XI})
\end{aligned} \tag{8.10}$$

Here α_t is the Lagrangian parameter for export and β_t the Lagrangian parameter for import.

The necessary first-order conditions are:

$$\frac{\partial L}{\partial e_t^H} = p_t'(e_t^H - e_t^{XI}) \cdot (e_t^H - e_t^{XI}) + p_t(e_t^H - e_t^{XI}) - \lambda \leq 0$$

$$(= 0 \text{ for } e_t^H > 0)$$

$$\frac{\partial L}{\partial e_t^{XI}} = -p_t'(e_t^H - e_t^{XI}) \cdot (e_t^H - e_t^{XI}) - p_t(e_t^H - e_t^{XI}) - \alpha_t + \beta_t + p_t^{XI} = 0$$

$$\lambda \geq 0 \; (= 0 \text{ for} \sum_{t=1}^{T} e_t^H < W) \tag{8.11}$$

$$\alpha_t \geq 0 \; (= 0 \text{ for } e_t^{XI} < \bar{e}^{XI})$$

$$\beta_t \geq 0 \; (= 0 \text{ for } -e_t^{XI} < \bar{e}^{XI})$$

We maintain the assumptions that the amount of electricity consumed at home, x_t, is positive in all periods and that the export/import prices are all different. Looking at the second condition, because we have either import or export in a period, the shadow prices on the upper and lower constraint cannot both be positive at the same time, but they may both be zero if the constraints are not binding.

We have by assumption that in an export period we must also use hydro at home. Therefore in an export period the flexibility-corrected price is also equal to the shadow price on water. The second condition in (8.11) tells us that the flexibility-corrected price is equal to the export price minus the shadow price on the export constraint. It will be arbitrary if export in each period of export is exactly equal to the constraint. In general there will therefore be a period when the export possibility is not fully utilised. We will call this period the *marginal export period* (see Chapter 6). But in this period the shadow price on water is equal to the export price. Denoting the period when the marginal export period occurs for t^* we have:

$$p_{t^*}(1 + \breve{\eta}_{t^*}) = \lambda = p_{t^*}^{XI} - \alpha_{t^*} = p_{t^*}^{XI} \tag{8.12}$$

But the shadow price on the water resource is a scalar. It is therefore the marginal export period that determines this shadow price. For all the export periods with a binding constraint the shadow prices on the upper constraint come in positive, satisfying the second equality in (8.11) for a general t belonging to the export periods (i.e., the periods when the export price is higher than the price for the marginal export period). The shadow prices are determined such that the difference between export price and the corresponding shadow price is constant and equal to the shadow price on water.

If hydro is used in an import period then the first condition in (8.11) holds with equality, implying that the flexibility-corrected home market price $p_t(1+\breve{\eta}_t)$ is equal to the shadow price on water. The second condition tells us that the flexibility-corrected price is always equal to the import price plus the shadow price on the upper constraint on import, yielding:

$$p_t(1+\breve{\eta}_t) = \lambda = p_t^{XI} + \beta_t \tag{8.13}$$

But by assumption $p_{t^*}^{XI} > p_t^{XI}$ for all periods being import periods. This means that hydro cannot be used in the home market in import periods unless the total import capacity is used. If hydro is not used in import periods the flexibility-corrected price is in a regular case lower than the shadow value on water and the import price is lower than the shadow value of water.

An illustration is provided in Figure 8.5. Because the import price is lowest in period 1, this period will be the import period. The original bathtub wall on the right-hand side is drawn with solid line, and on the left-hand side with a broken line. Both import and export capacities will be fully utilised. Because the import/export price is lowest in period 1, this will be the import period. The import capacity is added to the broken hydro wall to the left and marked with the solid vertical line. The demand and marginal revenue curves are anchored on the "import wall" on the left.

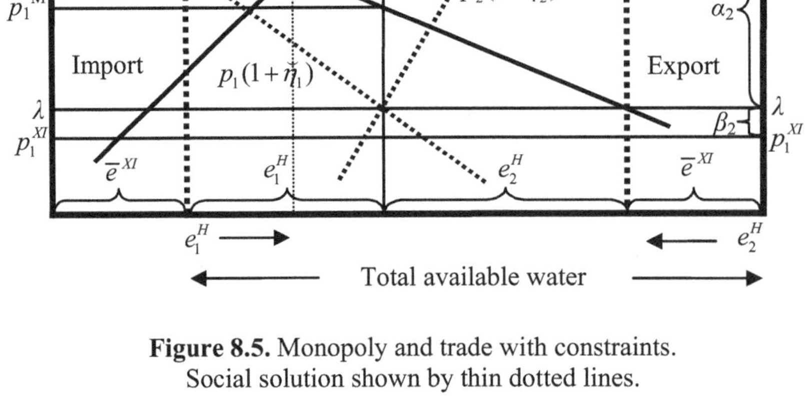

Figure 8.5. Monopoly and trade with constraints.
Social solution shown by thin dotted lines.

In the export period 2 the hydro wall on the right-hand side relevant for home consumption is shifted to the left with the length of the export constraint, marked with the broken, vertical line to the left of the right-hand hydro wall. This amount will be exported. The demand and marginal revenue curves relevant for the home country in period 2 are anchored on the broken, vertical wall. The flexibility-corrected prices are equal and equal to the shadow price on water. The home price becomes higher than the export price in the export period, and the home price becomes higher than the import price in the import period. The connection between the shadow price on water, the import/export prices, and the shadow prices on the trade constraints are shown in the figure.

Comparing with the social solution we have that both import and export trade capacity will be fully utilised, but that the home price will be equal for the two periods indicated by the dotted horizontal line through the point of intersection between the demand curves for the two periods. The monopolist will use more water at home in the relatively more price-elastic demand period 1 and accept a lower price than for the social solution (but higher than the import price), but then having less water left for the relatively inelastic period he will realise a higher price than both the social price and the export price.

Monopoly with reservoir constraints

Limited transferability of water between periods is the most realistic situation for hydropower. An upper limit on the reservoir will be introduced together with an accompanying water-accumulation equation. The monopoly problem is now based on the model (3.3) in Chapter 3 without trade possibilities. The profit maximisation problem is:

$$\max \sum_{t=1}^{T} p_t(e_t^H)\cdot e_t^H$$

subject to

$$\begin{aligned} & R_t \le R_{t-1} + w_t - e_t^H \\ & R_t \le \bar{R} \\ & R_t, e_t^H \ge 0\ ,\ t = 1,..,T \\ & T, w_t, R_o, \bar{R} \text{ given}, R_T \text{ free} \end{aligned} \tag{8.14}$$

The Lagrangian is:

$$
\begin{aligned}
L = &\sum_{t=1}^{T} p_t(e_t^H)e_t^H \\
&-\sum_{t=1}^{T} \lambda_t (R_t - R_{t-1} - w_t + e_t^H) \\
&-\sum_{t=1}^{T} \gamma_t (R_t - \bar{R})
\end{aligned}
\tag{8.15}
$$

The necessary first-order conditions are:

$$
\begin{aligned}
&\frac{\partial L}{\partial e_t^H} = p_t'(e_t^H)e_t^H + p_t(e_t^H) - \lambda_t \le 0 \ (= 0 \text{ for } e_t^H > 0) \\
&\frac{\partial L}{\partial R_t} = -\lambda_t + \lambda_{t+1} - \gamma_t \le 0 \ (= 0 \text{ for } R_t > 0) \\
&\lambda_t \ge 0 \ (0 = \text{for } R_t < R_{t-1} + w_t - e_t^H) \\
&\gamma_t \ge 0 \ (0 = \text{for}\, R_t < \bar{R}) , \qquad t = 1,..,T
\end{aligned}
\tag{8.16}
$$

Assuming electricity is always supplied and introducing the demand flexibility, $\breve{\eta}_t = p_t' e_t^H / p_t$, the first-order conditions read:

$$
\begin{aligned}
&p_t'(e_t^H)e_t^H + p_t(e_t^H) - \lambda_t = p_t(e_t^H)(1 + \breve{\eta}_t) - \lambda_t = 0 \\
&-\lambda_t + \lambda_{t+1} - \gamma_t \le \ (= 0 \text{ for } R_t > 0), \qquad t = 1,..,T
\end{aligned}
\tag{8.17}
$$

Comparing with the solution (3.5) of the social planning problem, the marginal revenue is substituted for the marginal willingness to pay (the price). The flexibility-corrected price is set equal to the water value, but the water values are period-specific, so marginal revenue may now differ over time. The second condition in (8.16) or (8.17), showing the dynamics of the water value, is qualitatively the same as in the social planning case. The discussion of the development of the water value is therefore qualitatively parallel to the social optimum case. By backward induction we can find the path of development for the water value. A general feature is that if the reservoir neither is threatened with overflow nor runs empty, the water value will remain constant and equal to the value in the terminal period. But in the monopoly case the market prices may fluctuate from period to period depending on changing demand functions.

In the social planning case discussed in Chapter 3 a quite reasonable assumption of non-satiation of electricity led to the terminal water value being positive in the case of free terminal reservoir level. In the monopoly case this assumption does not help us in general to determine the terminal value of the water value. To assume that the flexibility-corrected price for the terminal period will always be positive is a stronger assumption than assuming a positive price, or non-satiation. Such an assumption would imply that the monopolist will want to use up all available water in the terminal period.

The case of the terminal water value becoming zero does not create any formal problem. For a start it means that some water may be unused in the terminal period. If the upper reservoir constraint is not binding in the preceding period $T-1$ the water value will also be zero in this period, implying that the flexibility-corrected price is zero and water may be added to the reservoir handed to the terminal period. The water value can become positive only if there is a period where it is optimal to use up all available water. If this period is t, then we have from (8.17) that $\lambda_t \geq \lambda_{t+1} = 0$. The regular case will be that the water value for period t becomes positive. In the opposite case of a full reservoir in a period where all the later periods have zero water values, the water value cannot become less than zero. The shadow price on the upper constraint is in this case zero. Nothing is gained by expanding the reservoir limit marginally.

Introducing a lower limit on the terminal reservoir level or a scrap-value function as in (3.8) in the terminal period does not change the possibility of starting with a zero terminal water value when doing backwards induction. In the case of a lower positive constraint the monopolist may find it optimal to hand over more than this to the future, thus implying a zero shadow price on the terminal level. Using a scrap-value function, $S(R_T)$, following (3.9b), the condition for the terminal period becomes:

$$S'(R_T) - \lambda_T - \gamma_T = 0 \tag{8.18}$$

However, we cannot now exclude the possibility that the monopolist finds it optimal to deliver the maximal amount to the future in order to contract water usage within his planning period and even have overflow. The shadow price on the reservoir constraint becomes positive, because with a bigger reservoir more can be handed to the future contributing positively to the objective function. But the shadow price then becomes equal to the marginal value of the reservoir handed over, implying that the terminal water value in (8.18) is zero. In order to make sure to have a positive water value in the terminal period we have to assume that the marginal scrap value is higher than the shadow price on the reservoir constraint, implying no overflow. If the reservoir is not full in period $T-1$ the terminal water value will also be the water value for the preceding period. The discussion of possible water value developments will now parallel the discussion in Chapter 3 with flexibility-corrected prices substituting for social prices.

The general strategy of the monopolist of shifting water use from relatively inelastic demand periods to relatively elastic ones will also prevail in the case of a reservoir constraint. Let us first assume that the monopolist will not find it profitable to spill any water, i.e., that the marginal revenues stay positive. The constraint on the reservoir capacity will in general lead to the monopoly prices being closer to the prices in the social solution if the constraint is binding in the latter case. If it is optimal for a monopolist to have the upper constraint on the reservoir binding in a period, then this means that he must charge the market price given by the intersection of the demand curve and the vertical reservoir constraint in order to sell the available water. If the same amount of water is available as in the social case then the monopoly price must be equal to the price in the social optimum. The shadow value of water must adjust downwards for this to be possible. The monopolist follows the general strategy of using more water in elastic periods and having less water for the more inelastic periods. How this strategy interacts with storing more or less water than in the social planning case is connected to whether the reservoir build-up periods and the draw-down periods coincide with relatively elastic or inelastic periods. If build-up periods coincide with relatively elastic demand periods there will be a tendency to reduce the number of periods with binding reservoir constraint. Maximal storing may become more seldom the optimal strategy for a monopolist.

In the two-period illustration in Figure 8.6 the available water, including inflow and initial filling, in period 1 is *AC* and the inflow in period 2 is *CD*. The reservoir capacity is *BC*. The build-up period is period 1 with the most elastic demand. The reservoir constraint is not binding in the monopoly case, but was binding in the social optimal solution, as indicated by the dotted horizontal price lines intersecting the vertical reservoir constraint from *B*, and we have no spillage. The allocation point for water is moved from *B* in the social case to M^M in the monopoly case. We note that the monopoly price in period 1 with the relatively most elastic demand becomes lower than the social optimal price with a binding reservoir constraint, and the monopoly price in period 2 with relatively inelastic demand becomes higher than in the social optimal case. This is the general effect of shifting of water from periods with relative inelastic demand to periods with relatively elastic demand in the case of market power. The areas representing reduced income in period 1 and increased income in period 2 can easily be identified in Figure 8.6. Notice that the price differences are now quite reduced compared with the case of no reservoir constraint.

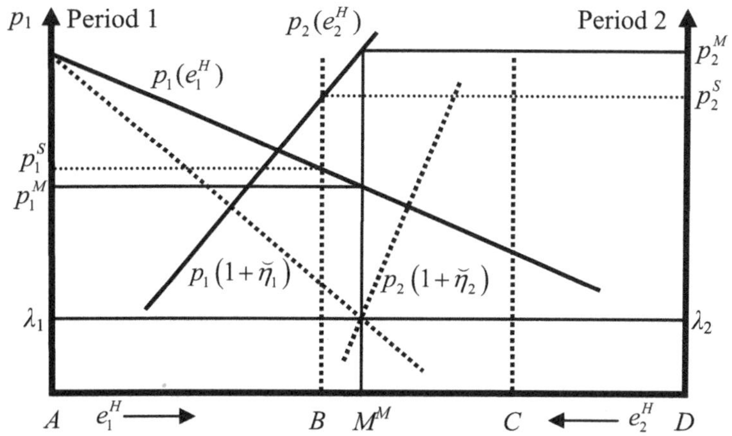

Figure 8.6. Monopoly with reservoir constraint.
Social solution shown by horizontal dotted lines.

It is often assumed that high demand periods, e.g., peak periods, are the periods with relatively most inelastic demand (Borenstein et al., 2002). However, this is an empirical question and should not be assumed without further investigations. Also in peak-demand periods there are substitution possibilities for consumers as pointed out in Chapter 1. In a summer period without both heating and cooling the substitution possibilities are much more restricted than in wintertime with several heating options, so it may as well be such periods that have the most inelastic demand as peak demand periods. The monopolist is utilising differences in demand elasticities and not differences in absolute demand.

A monopolist will experience a binding reservoir constraint as in the social case illustrated in Figure 8.6 if the intersection of marginal revenue curves is to the left of the vertical from *B* representing the reservoir constraint (the demand curves have to be slightly redrawn to obtain this case). In this case, if the monopolist tries to shift more water from inelastic periods to elastic periods, he will not maximise profits. In a two-period case with the same availability of water in the first period with the binding reservoir constraint the monopolist cannot do better than adopt the social solution although the demand in period 1 is more elastic.

Spilling of water can take place only in a period when the reservoir is filled up to the limit. The spilling then occurs if marginal revenue becomes zero before all available water in addition to the full reservoir is processed. Figure 8.7 illustrates such a case for the build-up period 1 having a less elastic demand than the draw-down period 2. The symbols have otherwise

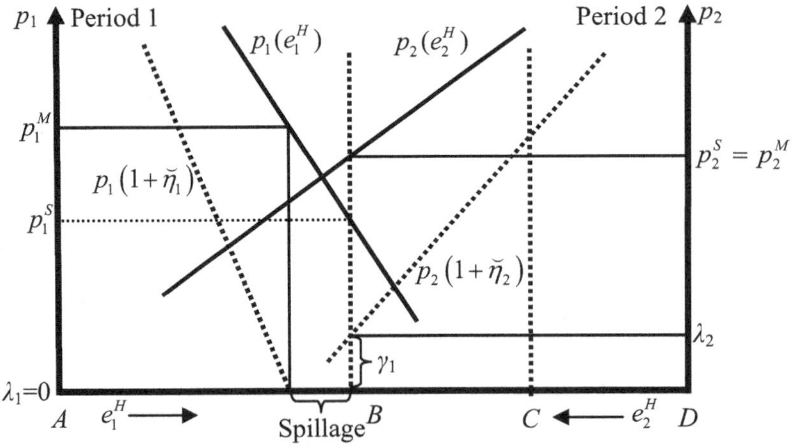

Figure 8.7. Monopoly with reservoir constraint and spillage. Social solution shown by dotted horizontal line.

identical interpretations with Figure 8.6. The marginal revenue becomes zero before all available water *AB* in addition to a full reservoir *BC* is processed, resulting in a spillage in period 1. The water value becomes zero according to the second condition in (8.17). The monopoly price is markedly increased compared with the social planning price, indicated by the thin horizontal dotted line from the intersection point between the demand curve for period 1 and the thick vertical broken line from *B* being the reservoir wall. However, because the marginal revenue curve for period 2 is hitting the reservoir wall at a positive value the monopolist will utilise all available water in period 2, implying he has to charge the same price as in the social planning case. There is a positive value of the shadow price on the reservoir constraint in period 1 equal to the differences between the water values for the two periods. Because the water value for period 1 is zero due to the overflow, the shadow price on the reservoir constraint become equal to the water value in period 2. If the reservoir could be expanded the monopolist will increase his profit with this amount at the margin. If period 2 is a peak period we see that the monopolist is not increasing the price in this period, but in the off-peak period because this period is relatively more inelastic.

Monopoly with trade and reservoir constraints

We will now combine trade and restriction on the reservoir. The monopoly optimisation problem in the case of restrictions both on trade and reservoir is:

$$\begin{aligned}
&\max \sum_{t=1}^{T} (p_t(x_t)x_t + p_t^{XI} e_t^{XI}) \\
&\text{subject to} \\
&x_t = e_t^H - e_t^{XI} \\
&-\bar{e}^{XI} \le e_t^{XI} \le \bar{e}^{XI}, \\
&R_t \le R_{t-1} + w_t - e_t^H \\
&R_t \le \bar{R} \\
&x_t, e_t^H, R_t \ge 0 \\
&T, \bar{R}, \bar{e}^{XI}, p_t^{XI} \text{ given, } e_t^{XI} \text{ free, } t = 1,..,T
\end{aligned} \tag{8.19}$$

Inserting the energy balance that holds as an equality constraint yields the Lagrangian:

$$\begin{aligned}
L = &\sum_{t=1}^{T} p_t(e_t^H - e_t^{XI}) \cdot (e_t^H - e_t^{XI}) + p_t^{XI} e_t^{XI} \\
&- \sum_{t=1}^{T} \lambda_t (R_t - R_{t-1} - w_t + e_t^H) \\
&- \sum_{t=1}^{T} \gamma_t (R_t - \bar{R}) \\
&- \sum_{t=1}^{T} \alpha_t (e_t^{XI} - \bar{e}^{XI}) \\
&- \sum_{t=1}^{T} \beta_t (-e_t^{XI} - \bar{e}^{XI})
\end{aligned} \tag{8.20}$$

The necessary first-order conditions are:

$$\frac{\partial L}{\partial e_t^H} = p_t'(e_t^H - e_t^{XI})(e_t^H - e_t^{XI}) + p_t(e_t^H - e_t^{XI}) - \lambda_t \le 0$$
$$(= 0 \text{ for } e_t^H > 0)$$

$$\frac{\partial L}{\partial e_t^{XI}} = -p_t'(e_t^H - e_t^{XI})(e_t^H - e_t^{XI}) - p_t(e_t^H - e_t^{XI}) - \alpha_t + \beta_t + p_t^{XI} = 0$$
$$\frac{\partial L}{\partial R_t} = -\lambda_t + \lambda_{t+1} - \gamma_t \le 0 \ (= 0 \text{ for } R_t > 0) \tag{8.21}$$
$$\lambda_t \ge 0 \ (= 0 \text{ for } R_t < R_{t-1} + w_t - e_t^H)$$
$$\gamma_t \ge 0 \ (= 0 \text{ for } R_t < \bar{R})$$
$$\alpha_t \ge 0 \ (= 0 \text{ for } e_t^{XI} < \bar{e}^{XI})$$
$$\beta_t \ge 0 \ (= 0 \text{ for } e_t^{XI} < -\bar{e}^{XI})$$

The change from the case of trade without reservoir restriction is that the water values are now period specific. Two consecutive water values are connected through the value of the shadow price on the reservoir constraint, as seen from the third condition in (8.21). The possibility of overflow may restrict import of electricity because water is used until the marginal revenue becomes zero if that is necessary to avoid overflow. In export periods home price may be driven further up because there is a limit on the transfer from the previous period. If the reservoir constraint does not become binding we are back to the solution without a reservoir constraint.

A bathtub illustration for two periods is provided in Figure 8.8, which is based on Figure 8.5. Because the import price is lowest in period 1 this period will again be the import period. Available water including inflow to the reservoir in period 1 is *AC* and inflow in period 2 is *CD*. The size of the reservoir is *BC*, indicated by $\bar{R}$, and the broken, vertical lines from *B* and *C* represent the reservoir. The reservoir is introduced from *C* to the left to *B* because our problem for two periods is how much water to leave to period 2. The import constraint, indicated by the solidly drawn energy wall, is placed to the left of the hydropower wall, drawn with a broken line from *A*. In our case the full import capacity will not be utilised. But the full export capacity will be used, and this capacity is indicated by the first thick, dotted line to the left of the right-hand hydro wall drawn with a solid line.

The final layout of the figure may be thought of as the result of two stages, where only the last stage is drawn, for the two periods' curves. In the first stage the demand and marginal revenue curves are anchored to the hydropower walls erected from *A* and *D*. The optimality conditions for the import period tell us that the marginal revenue curve should pass through the intersection between the import price line and the hydro wall from *B*.

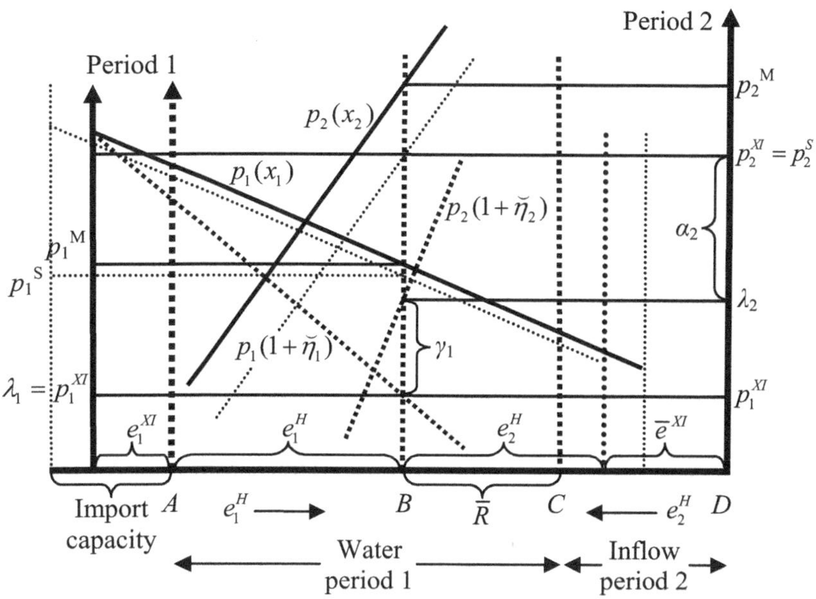

Figure 8.8. Monopoly, trade, and reservoir constraints. Social solution shown by thin dotted lines.

The demand and marginal revenue curves are then shifted horizontally to the left to allow this, and the stopping point is where the import wall is erected. If more import is tried the marginal revenue will become smaller than the import price. At least water AB has to be used home in period 1, and the market price matching this amount is higher than the import price. Therefore import is introduced until the marginal revenue is equal to the import price. Recall the analogy between imports and another technology for producing electricity. The final market price is found the usual way of moving vertically up to the demand curve. Because the import capacity is not fully utilised the shadow price β_I on this capacity is zero. The water value becomes equal to the import price for this period. The maximal amount of water BC is transferred to period 2. Checking period 2, there is in the first stage enough water to utilise the export capacity fully. The thick, vertical dotted line to the left of the hydropower wall then indicates the reduced availability for hydropower at home, and the demand and marginal revenue curve are shifted horizontally to the left and anchored to the new wall. The intersection of the vertical water storage wall from B and the marginal revenue curve for period 2 then gives the water value for period 2. The home price is found by the intersection of the hydropower storage line and the demand curve. The shadow price γ_1 on the reservoir capacity is the difference between the two periods' water values and is indicated in the figure. Since the export capacity is fully utilised its shadow price α_2 is positive and indicated as the difference between the export price and the water value for period 2.

Entering thin dotted lines for the solution of the social-planning case facilitates a comparison with the monopoly case. The import and export periods remain the same. The import capacity will now be fully utilised, so the demand curve for period 1 will be anchored at this import-extended wall, illustrated by the thin dotted vertical line to the left of the bathtub wall in the monopoly case. In addition, all water that cannot be transferred to period 2 will be used at home in the import period, resulting in slightly more use of water in the social case in period 1 and a slightly lower price than in the monopoly case. In period 2 the full export capacity will not be used because using it will leave so little water to be consumed at home that the market price will increase above the exogenous export price. Only such an amount will be exported that lead to the same price at home as the export price. The demand curve for period 2 must therefore pass through the intersection point of the export price line and the broken storage wall erected from *B*. The demand curve is anchored (not shown in the figure) at the thin vertical dotted line to the right of the monopoly anchoring indicating the reduced optimal export in the social case. In our illustration monopoly leads to a shift away from imports and over to exports. Because import is reduced the monopoly price is (slightly) higher in the import period. Because the same total amount of water is transferred to period 2 in the monopoly case the increased export leads to a (markedly) higher domestic price and a reduced consumption. The export period has the relatively most inelastic demand.

Monopoly with hydro and thermal plants

Hydro is in most countries combined with thermal capacity. Let us first assume that a monopolist has full control over both hydro and thermal capacity. The thermal capacity is aggregated into a sector capacity by using an aggregate merit-order cost function as explained in Chapter 5. We will investigate how the monopolist utilises the two types of electricity technologies compared with the social solution. We assume that the monopolist is free to reduce production e_t^{Th} from the thermal units as he sees in his interest. The simplest restriction on hydro production of a total available amount of water is used. Thermal capacity is restricted to $\bar{e}^{Th}$. The demand functions are $p(x)$, where x is the electricity demand supplied both by hydro and thermal capacity. The optimisation problem, adapted from (5.15) is:

$$\begin{aligned}
&\max \sum_{t=1}^{T}[(p_t(x_t)x_t - c(e_t^{Th})] \\
&\text{subject to} \\
&x_t = e_t^H + e_t^{Th} \\
&\sum_{t=1}^{T} e_t^H \le W \\
&e_t^{Th} \le \bar{e}^{Th} \\
&x_t, e_t^H, e_t^{Th} \ge 0, \;\; t = 1,..,T \\
&T, W, \bar{e}^{Th} \text{ given}
\end{aligned} \tag{8.22}$$

Substituting for total energy the Lagrangian is

$$\begin{aligned}
L = &\sum_{t=1}^{T}[p_t(e_t^H + e_t^{Th})(e_t^H + e_t^{Th}) - c(e_t^{Th})] \\
&-\sum_{t=1}^{T}\theta_t(e_t^{Th} - \bar{e}^{Th}) \\
&-\lambda(\sum_{t=1}^{T} e_t^H - W)
\end{aligned} \tag{8.23}$$

The necessary conditions are:

$$\begin{aligned}
&\frac{\partial L}{\partial e_t^H} = p_t'(e_t^H + e_t^{Th})(e_t^H + e_t^{Th}) + p_t(e_t^H + e_t^{Th}) - \lambda \le 0 \\
&\quad (= 0 \text{ for } e_t^H > 0) \\
&\frac{\partial L}{\partial e_t^{Th}} = p_t'(e_t^H + e_t^{Th})(e_t^H + e_t^{Th}) + p_t(e_t^H + e_t^{Th}) - c'(e_t^{Th}) - \theta_t \le 0 \\
&\quad (= 0 \text{ for } e_t^{Th} > 0) \\
&\lambda \ge 0 \; (= 0 \text{ for } \sum_{t=1}^{T} e_t^H < W) \\
&\theta_t \ge 0 \; (= 0 \text{ for } e_t^{Th} < \bar{e}^{Th})
\end{aligned} \tag{8.24}$$

Concentrating on periods where both hydro and thermal are used, the general result is that marginal revenue substitutes for the marginal willingness to pay in the social optimal solution:

$$p_t(x_t)(1+\breve{\eta}_t) = \lambda = c'(e_t^{Th}) + \theta_t \qquad (8.25)$$

The monopoly solution for a period is illustrated in Figure 8.9. If the monopolist's water value is *OB* in a period, total energy supplied is indicated by the intersection of the horizontal water value line *BB'* and the marginal revenue curve, yielding quantity Oe^H and monopoly price p^M. Both thermal and hydro capacity will be used according to the marginal revenue condition (8.25). The thermal capacity will be Oe^{Th}, determined by the intersection between the marginal cost curve and the water value line *BB'* at *b*, and the hydro capacity ($Oe^H - Oe^{Th}$). The thermal capacity is not exhausted, so the shadow price on thermal capacity is zero.

For two periods we may again use the bathtub diagram to illustrate the allocation of the two types of power on the two periods. In Figure 8.10 the length of the hydro bathtub, *BD*, is extended at each end with the thermal capacity. The thermal marginal cost functions are anchored at the hydro walls and extending to the left out to the capacity limit indicated by a short vertical line for period 1 and to the right for period 2, as explained in Chapter 6. Using the result (8.25), with the shadow price on the thermal capacity constraint being zero, we have that the thermal extension of the bathtub is equal at each end; with *AB* in period 1 and *DE* in period 2 and *AB* = *DE*. The equilibrium allocation is at point *C*, resulting in an allocation of *AB* thermal and *BC* hydro in period 1, and *CD* hydro and *DE* thermal in period 2. Although all available water may be used in both periods

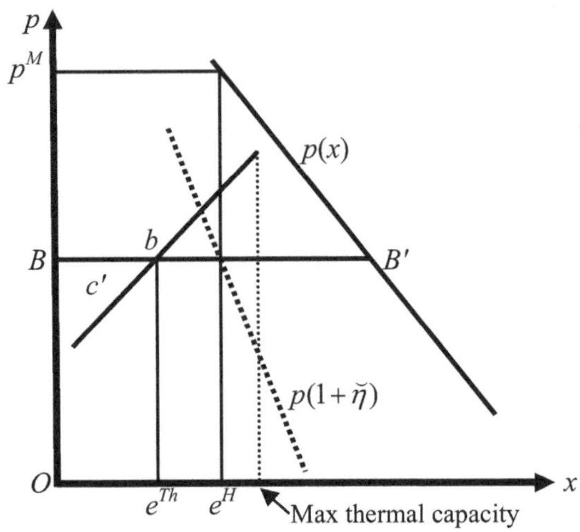

Figure 8.9. Monopoly. Hydro and thermal capacity.

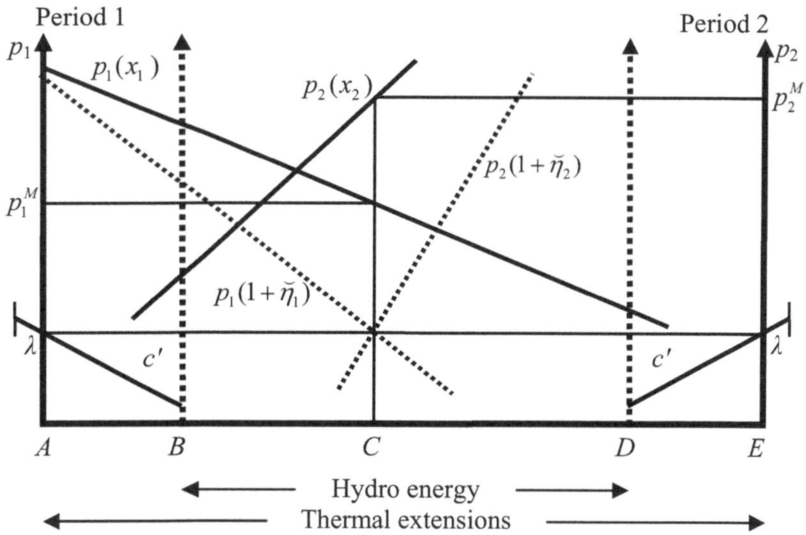

Figure 8.10. Two periods and monopoly with hydro and thermal capacity.

as is the case in Figure 8.10, the monopolist will reduce the use of thermal capacity. This may be seen recalling that the water value in the monopoly case will always be lower than the optimal social price. Since the monopolist equates his water value with the marginal cost of thermal capacity, the result follows.

Introducing a reservoir constraint as in Figure 8.6 will not change the solution for the case of an intersection of the marginal revenue curves within the area delimited with the lines from B and C in that figure showing the storage possibilities. A monopolist will equate the water value with the marginal cost of thermal, and not the market price. Compared with the social-planning solution the use of thermal capacity may be reduced in all periods and will be base load unless a hydro reservoir constraint is binding. For such periods thermal capacity will also be used as peak.

Dominant firm with a competitive fringe

A pure monopoly in the electricity market is not so common. There may be a dominating firm in terms of market share, but there will often be many smaller firms acting as price takers in the market. The existence of such a competitive fringe reduces the possibility of using market power because the fringe firms will supply according to the market price. For simplicity we will model the dominant firm by using the hydro model (8.1) without a reservoir constraint, but with a total water constraint, and model the competitive fringe by introducing a thermal sector represented by a cost function, as in the previous section, but without imposing a capacity constraint for the time being.

The optimisation problem for the dominating hydro producer is:

$$\begin{aligned}
&\max \sum_{t=1}^{T} p_t(x_t)e_t^H \\
&\text{subject to} \\
&x_t = e_t^H + e_t^{Th} \\
&\sum_{t=1}^{T} e_t^H \le W \\
&p_t(x_t) = c'(e_t^{Th}) \\
&x_t, e_t^H, e_t^{Th} \ge 0,\ \ t = 1,..,T \\
&T, W \text{ given}
\end{aligned} \tag{8.26}$$

The third constraint in (8.26) represents the behaviour of the competitive fringe. It supplies according to the price-taking profit maximising condition of equating market price with marginal costs. We can most conveniently proceed in the standard textbook way by using the third condition to derive the relationship between the supply of the fringe and the dominant producer's supply of hydroelectricity. If the hydro producer supplies more the market price *cet. par.* goes down, but then the fringe contracts its output, assuming that the marginal cost is increasing. Differentiating

$$p_t(e_t^H + e_t^{Th}) = c'(e_t^{Th})(t = 1,..,T) \tag{8.27}$$

yields:

$$\begin{aligned}
&p_t'(e_t^H + e_t^{Th})(de_t^H + de_t^{Th}) = c''(e_t^{Th})de_t^{Th} \Rightarrow \\
&\frac{de_t^{Th}}{de_t^H} = \frac{-p_t'(e_t^H + e_t^{Th})}{p_t'(e_t^H + e_t^{Th}) - c''(e_t^{Th})} < 0\ \ (t = 1,..,T)
\end{aligned} \tag{8.28}$$

Equation (8.27) defines implicitly the fringe output as a function of the output of the dominant firm. The relationship can be expressed by

$$e_t^{Th} = f_t(e_t^H), f_t' < 0\ (t = 1,..,T) \tag{8.29}$$

Using the energy balance and the relationship between fringe output and output of the dominating firm yields a more compact problem than (8.26) with the Lagrangian as

$$L = \sum_{t=1}^{T} p_t(e_t^H + f_t(e_t^H))e_t^H$$
$$-\lambda(\sum_{t=1}^{T} e_t^H - W) \tag{8.30}$$

The first-order conditions are:

$$\frac{\partial L}{\partial e_t^H} = p_t(e_t^H + e_t^{Th}) + p_t'(e_t^H + e_t^{Th})e_t^H(1+\frac{de_t^{Th}}{de_t^H}) - \lambda \le 0$$
$$(= 0 \text{ for } e_t^H > 0) \tag{8.31}$$
$$\lambda \ge 0\ (= 0 \text{ for } \sum_{t=1}^{T} e_t^H < W)\ ,\quad t = 1,..,T$$

The last bracketed term, $(1 + de_t^{Th}/de_t^H)$, on the right-hand side of the first condition in (8.31) is positive, but less than 1, resulting in the conditional marginal revenue becoming less than the price. Using (8.28) yields:

$$1+\frac{de_t^{Th}}{de_t^H} = 1+\frac{-p_t'(e_t^H + e_t^{Th})}{p_t'(e_t^H + e_t^{Th}) - c''(e_t^{Th})} = \frac{-c''(e_t^{Th})}{p_t'(e_t^H + e_t^{Th}) - c''(e_t^{Th})} > 0 \tag{8.32}$$

The marginal revenue of the dominant firm is now reflecting the behaviour of the fringe. We have that the value of the conditional marginal revenue is closer to the market price for a given total quantity (but still below this value) compared with the expression for monopoly marginal revenue. Rearranging the first-order condition in (8.31) yields the following expression for the conditional marginal revenue:

$$MR_{t|p_t'=c'} = p_t(1+\breve{\eta}_t \frac{e_t^H}{e_t^H + e_t^{Th}}) + p_t' \frac{de_t^{Th}}{de_t^H} e_t^H\ ,\ t = 1,..,T \tag{8.33}$$

The conditional marginal revenue function is closer to the demand function than the monopolist's marginal revenue function because of two factors: the market share of the dominant firm is less than 1 in the first expression in (8.33) reducing the impact of the demand flexibility, and the second expression involving the quantity reaction of the fringe is positive.

When the dominant firm is producing (8.31) tells us that the marginal revenues conditional upon the behaviour of the fringe shall all be equal and equal to the shadow price on water. It seems reasonable to assume that the dominant firm produces in all periods. Zero production implies that the shadow value of water is greater than the marginal cost of the fringe providing the whole market quantity. We will disregard this possibility.

An illustration in the two-period case is provided by Figure 8.11. The broken lines below the demand curves are the conditional marginal revenue curves. The optimal solution is characterised by these conditional marginal revenues being equal and equal to the shadow price of water. The use of the fringe thermal capacity is governed by the equality of the market price and the marginal cost. The demand and conditional marginal revenue curves are anchored on the thermal walls, being endogenously determined, extending the energy bathtub like the case in Figure 8.10. The thermal cost functions are anchored on the hydro bathtub walls. The use of thermal capacity, *AB*, in the relatively elastic period 1 is smaller than the use *DE* in the more inelastic period 2. The market prices differ and the price is highest in the more inelastic period. Thus the existence of a fringe leads the dominant firm to use more thermal capacity in the high price period than in the low price period, in contrast to the monopoly case. If the relatively inelastic period is the peak period this means that thermal is now serving as peak capacity and not only as base load as in the monopoly case. In the illustration more hydro, *BC*, is used in period 1 than in period 2, using *CD*. Compared with the monopoly case the impact of the fringe is clearly to

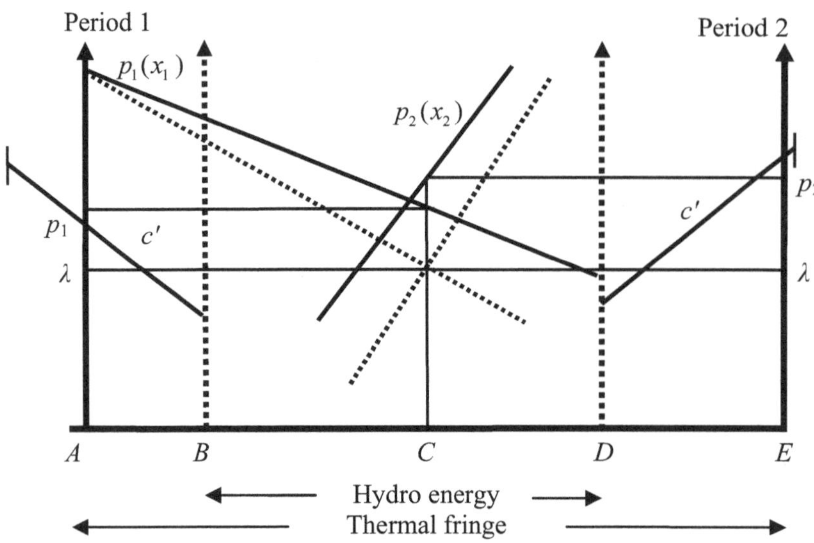

Figure 8.11. Dominant hydro and a thermal fringe.

make the prices become more equal. A larger fringe capacity will be used in the more inelastic period, forcing the market price down. This reduces the effectiveness of shifting water from period 2 to period 1. Using more water in period 1 is actually more effective *cet. par.* for the dominant firm in the sense that the necessary price decrease is cushioned because the fringe will contract its output. However, the fringe activates more capacity in the high price period; thus the existence of a fringe leads to less market power being exercised.

A constraint on the thermal capacity of the fringe will be an advantage for the dominant hydro firm if the constraint becomes binding. The first-order profit-maximising conditions for the price-taking fringe in the case of a capacity constraint are:

$$\begin{aligned} p_t(x_t) &= c_t'(e_t^{Th}) + \theta_t \\ \theta_t &\geq 0 \; (= 0 \text{ for } e_t^{Th} < \overline{e}^{Th}) \end{aligned} \qquad (8.34)$$

The capacity constraint is $\overline{e}^{Th}$ and its shadow price θ_t. The capacity restriction implies the following response of the fringe:

$$e_t^{Th} = \overline{e}^{Th} \text{ for } p_t(x_t) \geq \overline{p} = c'(\overline{e}^{Th}) \qquad (8.35)$$

In the case of $p_t(x_t) \geq \overline{p}$ the first-order condition (8.31) for the dominant firm becomes

$$\begin{aligned} & p_t(e_t^H + \overline{e}^{Th}) + p_t'(e_t^H + \overline{e}^{Th})e_t^H = \\ & p_t(1 + \bar{\eta}_t \frac{e_t^H}{e_t^H + \overline{e}^{Th}}) = \lambda \,, \; t = 1,..,T \,, \end{aligned} \qquad (8.36)$$

assuming that the dominating firm is producing. The conditional marginal revenue function shifts further away from the demand function. But since the demand flexibility is multiplied with the market share of the dominating firm this implies that the conditional marginal revenue function does not shift down as far as to the monopoly marginal revenue function.

In the two-period case the situation can be illustrated as in Figure 8.12, building upon Figure 8.11. Total hydro resource is *BD*. The capacity of the thermal fringe is indicated by the small vertical line at the end of the marginal cost curve outside the thermal wall in period 1. The thermal capacity constraint is binding in period 2, but not in period 1. The demand and marginal revenue curves for period 2 are now anchored on the thermal wall dictated by the capacity constraint. The shift to the marginal revenue curve defined in (8.36) valid when the fringe output is constrained, is shown by the greater distance between the demand and the marginal revenue curve.

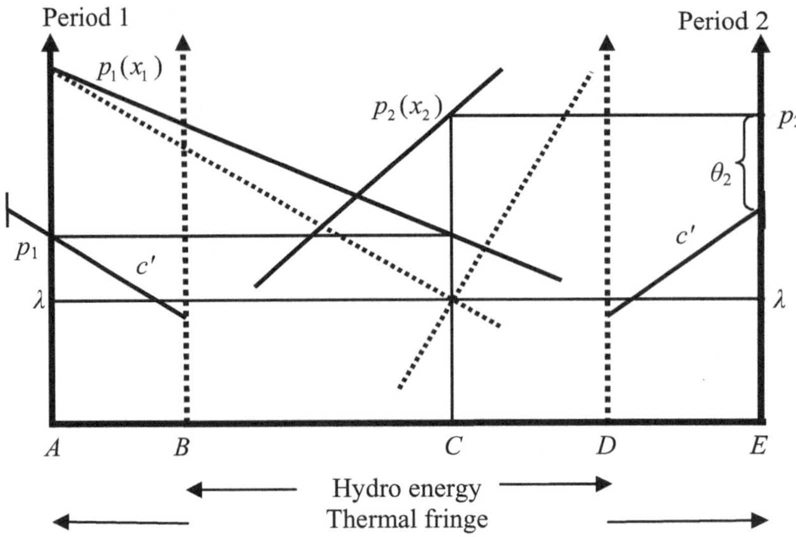

Figure 8.12. Dominant firm and constraint on the fringe output.

The opposite direction of shifts for the demand and revenue curves implies an increase in period 2 price. Total supply is *CE*, the fringe supplies its maximal capacity *DE*, and the dominant firm supplies *CD*. In period 1 the thermal capacity is not fully utilised and the conditional marginal revenue curve follows from (8.31) and lies relative closer to the demand curve as in Figure 8.11. The fringe supplies *AB*, less than its capacity, and the dominant firm supplies *BC*. When allocating water between the two periods the dominant hydro firm strikes a balance between marginal income from the two periods, taking into consideration the lack of quantity response from the fringe in period 2 with full capacity utilisation and the contracting response in period 1 if more water is shifted to this period. The shadow price θ_2 on the thermal capacity constraint in period 2 is shown in the figure and is the difference between the market price and the marginal cost at full capacity. Thus it measures the revenue to the fringe of expanding capacity marginally. The size of the capacity shadow price is also an indication for the dominant firm of the advantage enjoyed due to the fringe being capacity constrained.

Hydro producers can also constitute a fringe. However, the behaviour of the fringe can lead to analytical problems finding an optimal solution to the profit maximising problem of the dominating firm. Assuming that the fringe has at its disposal an amount of water corresponding to W_F and has enough reservoir capacity to be perfectly flexible as to in which period to use the water, the fringe will use all its water in the period with the highest price. Although it is a fringe and therefore W_F may be considerably smaller than W_D, where W_D is the dominant firm's water resource, it can still have a considerable market share if all its water is used just in one period. It may happen that for a relatively large fringe water resource the solution is forced to be the social optimal solution with equal price for all periods.

Introducing reservoir constraints for the fringe may introduce some market power for the dominant firm. But it is then also logical to introduce a reservoir constraint for the dominant firm. We will not develop such an analysis further, but just mention that in Norway the reservoir capacity is quite concentrated on a small number of firms. Small hydropower firms tend to have relatively less reservoir capacity, thus opening up for the possibility of a group of dominating firms to exercise some market power.

Oligopolistic markets

It may easily become difficult to analyse oligopolistic markets involving hydro producers analytically. The basic problem is that such analyses have to be dynamic due to the basic dynamic nature of optimal adjustments of hydro producers with reservoir capacity. As shown in Garcia et al. (2001) and Kelman et al. (2001), oligopoly models involving hydro producers require solving differential games. Even a Cournot duopoly involving a hydro firm and a thermal firm may become intractable without assuming special functional forms for the demand and cost functions considering only two periods (Crampes and Moreaux, 2001). Since there is zero variable cost in the hydro case Bertrand competition of moving prices is of special interest. A hydro producer can more easily drive down the price in the short run and force thermal capacity out and use water in order to create more scarcity in later periods. We do not attempt to develop such analyses here.

Chapter 9. Uncertainty

The general problem

A very basic feature of hydropower operation that has been neglected so far is that inflows to the reservoirs are stochastic variables. Weather is predicted, but as we all know with varying accuracy. The problems this creates for hydropower management are quite obvious. A decision about use of water, i.e., production in the current period and transferring water to the next period, has to be made in the current period while the inflows of the future periods up to the horizon are known only by their predictions. The best we can do in the current period is to formulate an optimal plan by maximising the *expectation* of the sum of consumer plus producer surpluses. The demand functions themselves may also be influenced by the weather. It is obvious that the need for both space heating and cooling depends on the outside temperature. But the temperature must also be regarded as a stochastic variable. Further real-life stochastic events in the case of a complete electricity system with transmission lines and thermal capacities are transmission capacity being reduced due to transformer accidents, storms blowing down trees on lines, breaking of lines due to icing, etc., and thermal capacities going down due to accidents. Considering windmills the output depends crucially on the wind speed that is stochastic.

The problem for finding optimal solutions of the hydro management problem created by uncertainty was recognised early in the literature (Little, 1955; Koopmans, 1957; Gessford and Karlin, 1958; Morlat, 1964[1]). In Norway a special solution strategy termed the expected water value approach was introduced in Hveding (1967-1968) based on Stage and Larsson (1961). In the more specialised engineering literature an early contribution was Pereira (1989).

[1] Morlat noted that he built his uncertainty analysis on Massé (1946).

Reformulating the most realistic model based on a set of hydropower plants with one reservoir each and upper constraints on reservoirs and production capacities, model (4.8) in Chapter 4, yields the social planning problem

$$\max \sum_{t=1}^{T} E\left\{\int_{z=0}^{x_t} p_t(z)\,dz\right\}$$

subject to

$$\begin{aligned}
& x_t = \sum_{j=i}^{N} e_{jt}^{H},\ t = 1,..,T \\
& R_{jt} \le R_{j,t-1} + w_{jt} - e_{jt}^{H} \\
& R_{jt} \le \overline{R}_j \\
& e_{jt}^{H} \le \overline{e}_j^{H} \\
& R_{jt}, w_{jt}, e_{jt}^{H} \ge 0 \\
& T, N, R_{jo}, \overline{R}_j, \overline{e}_j^{H} \text{ given, } R_{jT} \text{ free, } j = 1,..,N,\ t = 1,..,T
\end{aligned} \tag{9.1}$$

Parameters of the demand functions $p_t(x_t)$ ($t = 1,\ldots,T$) are stochastic variables. We may assume that their probability distributions are known, and that these distributions vary with the period, t. Since inflows are stochastic, so are the reservoir levels in the case of no threat of overflow and so are the production levels of each plant, e_{jt}^{H} ($j = 1,\ldots,N$), because they depend on the reservoir level of current and past period. For a realistic dimension of the problem, i.e., of the order of $3TN$, with T in the case of using a week as a period being in the order of 52 to 260 (five years) and N being over 700 in the case of Norway, this is not a trivial problem to solve. But taking care of the constraints in (9.1) by formulating a Lagrangian function as in (4.8) is no longer an appropriate procedure. The qualitative nature of the solution cannot be worked out starting with first-order conditions for a time period $t < T$. Stochastic variables appear in all conditions for periods $t + 1$ to T, and the Lagrangian parameters themselves will become stochastic variables. The only way of establishing the nature of the solution is to use Bellman's principle of backwards induction.

In the engineering literature problems like (9.1) are solved numerically using discrete-time stochastic dynamic programming formulations (Wallace and Fleten, 2002). Solution algorithms have been developed over the last decades in the engineering literature approximating optimal solutions (Pereira, 1989; Pereira and Pinto, 1985, 1991). However, even with modern computers the number of possible combinations of realisations of

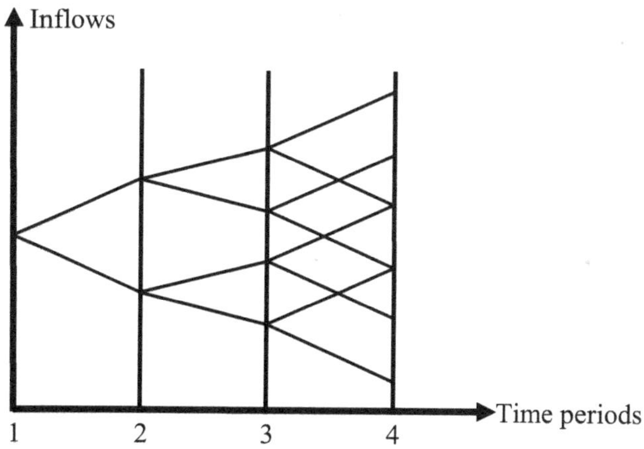

Figure 9.1. Possible realisations of inflows.

stochastic variables in real-life large-scale problems has been too much to allow global numerical optimal solutions to be found. Recently approaches based on dual stochastic dynamic programming seem to be promising. However, the solutions are not analytic, but have to be calculated numerically.

To see the futility in trying to solve the problem (9.1) starting with the first year we will consider inflows only as stochastic. The possible realisations of inflows can be illustrated with a familiar tree diagram as shown in Figure 9.1, showing only two alternatives for ease of exposition for the inflow values that may differ over the periods. Starting out from a certain inflow in period $t = 1$ the potential inflows for the other periods branch out and the actual development of inflows can follow quite different patterns over time, assuming that inflows in one period is independent of the inflow in the previous period. It is not feasible to solve the problem starting from the first period, but the approach of Bellman (1957) has to be followed, starting with the terminal period.

We will only be concerned with qualitative conclusions we can find about the nature of the optimal management solution and will therefore consider very simplified settings (Hansen, 2007).

A simplified two-period approach

In order to have a simple model, demand is assumed to be deterministic and only inflows to be stochastic. Furthermore, only the reservoir constraint will be introduced; the production constraint is dropped. Only an aggregated system consisting formally of one plant and one reservoir will be considered. The equation of motion of the reservoir level can be rewritten, taking into consideration that the reservoir level must be the minimum of the upper limit of the reservoir and what is accumulated during a period from the level that was transferred from the previous period:

$$R_t = \min\left[\bar{R}, R_{t-1} + w_t - e_t^H\right],\ R_t \geq 0, t = 1,..,T \tag{9.2}$$

The inflows w_t are stochastic with a known distribution that is period specific. Simplifying further by assuming that the inflow for the current period is known, the decision problem evaluated in period 1 under uncertainty becomes:

$$\begin{aligned} &\max\left[\int_{z=0}^{e_1^H} p_1(z)dz + \sum_{t=2}^{T} E\left\{\int_{z=0}^{e_t^H} p_t(z)dz\right\}\right] \\ &\text{subject to} \\ &R_t = \min\left[\bar{R}, R_{t-1} + w_t - e_t^H\right], \\ &R_t, w_t, e_t^H \geq 0,\ t = 1,..,T \\ &T, R_o, \bar{R}\ \text{given},\ R_T\ \text{free} \end{aligned} \tag{9.3}$$

Simplifying even further to just two periods allows us to express the production in period 2 as a function of only one stochastic variable, the inflow in period 2, and deterministic variables from period 1. Transforming the constraint in (9.3) yield the problem:

$$\begin{aligned} &\max\left[\int_{z=0}^{e_1^H} p_1(z)dz + E\left\{\int_{z=0}^{e_2^H} p_2(z)dz\right\}\right], \\ &e_2^H = \min\left[\bar{R} + w_2, R_o + w_1 + w_2 - e_1^H\right] \end{aligned} \tag{9.4}$$

The transfer of water from period 1 to period 2 is in general the smallest amount of $R_1 = R_o + w_1 - e_1^H$ and $R_1 = \bar{R}$. However, under our general assumption of non-satiation in every period it is quite intuitive that it cannot be optimal with overflow. The objective function for period 1 is increasing in electricity consumption. If it should be optimal that the maximal amount $\bar{R}$ is transferred to period 2 from period 1 we will have $R_1 = \bar{R} = R_o + w_1 - e_1^H$, i.e., no overflow in period 1. Formally, this can be demonstrated investigating the situation that $R_1 = \bar{R}$. Since this is the maximal amount that can be handed over to period 2 we can then look at period 1 only and ask what amount it is optimal to consume in period 1. This is found as the solution to the following problem:

$$\begin{aligned}
&\max \int_{z=0}^{e_1^H} p_1(z)\,dz \\
&\text{subject to} \qquad\qquad\qquad (9.5)\\
&e_1^H \le R_o + w_1 - \bar{R} \\
&R_o, \bar{R} \text{ given, } e_1^H \ge 0
\end{aligned}$$

The Lagrangian for the problem is:

$$L = \int_{z=0}^{e_1^H} p_1(z)\,dz - \mu(e_1^H - (R_o + w_1 - \bar{R})) \qquad (9.6)$$

The necessary first-order conditions are:

$$\begin{aligned}
&\frac{\partial L}{\partial e_1^H} = p_1(e_1^H) - \mu \le 0 (= 0 \text{ for } e_1^H > 0) \\
&\mu \ge 0 (= 0 \text{ for } e_1^H < R_o + w_1 - \bar{R})
\end{aligned} \qquad (9.7)$$

Appealing to realism we assume that there is positive electricity production in every period, i.e., $e_1^H > 0$. This implies that $p_1(e_1^H) > 0$ and $p_1(e_1^H) = \mu > 0$. From the complementary slackness condition in (9.7) we then have that $e_1^H = R_o + w_1 - \bar{R}$, i.e., the maximal amount that can be consumed in period 1 is consumed. There is no waste. The shadow price on the upper constraint is positive and determined as

$$\mu = p_1(R_o + w_1 - \bar{R}) = p_1^{\max} \qquad (9.8)$$

In period 1 it is known that in period 2 all available water will always be utilised since there is no requirement on the terminal value, and by assumption the marginal utility of electricity remains positive even for the maximal possible amount of water in period 2, i.e., $\bar{R}$ plus maximal amount of inflow that can occur in the known probability distribution for inflow in period 2. Therefore it is known in period 1 that it cannot be possible with threat of overflow in period 2.

A bathtub diagram, Figure 9.2, can be used to illustrate the situation. Production in period 1 is measured from the left-hand vertical axis and production in period 2 from the right-hand axis in the usual way. Seen from period 1 the placement of the right-hand axis is stochastic. Two extreme realisations from the distribution of inflow in period 2 are indicated; w_2^{min} and w_2^{max}. The minimum amount may be zero. In period 1 the amount $AC\ (= R_o + w_1)$ is available for production and for transfer to period 2. The size of the reservoir is measured from the point of the available water from right to left and is Bw_2^{min} for the lowest possible inflow and Cw_2^{max} for the highest possible inflow (both equal to $\overline{R}$), with the left-hand reservoir limits indicated by the vertical broken lines from B and C. The inflows in period 2 are in the figure measured from C, i.e., the minimum amount of inflow is in the diagram Cw_2^{min} and the maximum Cw_2^{max}. The demand function for period 2 shifts according to the realisation of the inflow and the curves corresponding to the minimal and maximal inflows are marked "min" and "max."

Problem (9.4) can be reformulated by substituting for the consumption in period 2 by using the second equation in (9.4), having established that it cannot be optimal with overflow in period 1, implying that the minimal optimal choice of consumption of electricity in period 1 is the level corresponding to AB, making the consumption in period 2 always equal to what is handed over from period 1 and the inflow in period 2; $e_2^H = R_o + w_1 + w_2 - e_1^H$. If the maximal amount of water is handed over, we have $\overline{R} = R_o + w_1 - e_1^H$ and the previous equality still holds. The optimal choice of e_1^H is restricted to the interval AC in the figure. The optimi-

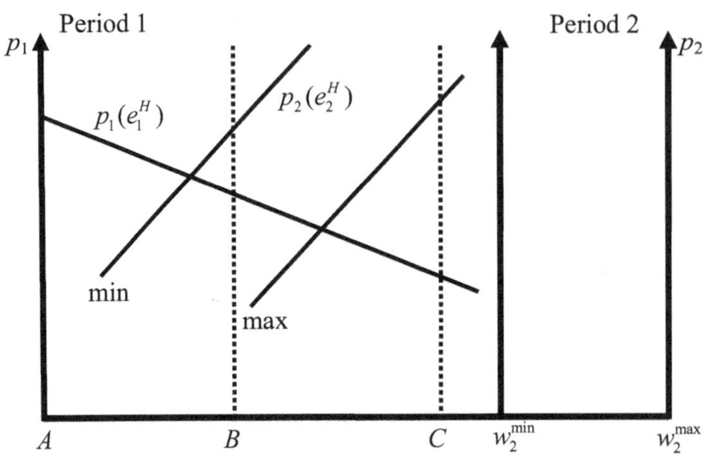

Figure 9.2. Stochastic inflows in period 2.

sation problem becomes:

$$\max_{e_1^H} \left[\int_{z=0}^{e_1^H} p_1(z)dz + E\left\{ \int_{z=0}^{R_o+w_1+w_2-e_1^H} p_2(z)dz \right\} \right]$$

$$\text{subject to} \qquad (9.9)$$

$$e_1^H \in \left[\max(0, R_o + w_1 - \bar{R}), R_o + w_1\right]$$

$$R_1 \in \left[0, \bar{R}\right]$$

The last reservoir constraint is not an independent constraint, but follows from the restrictions on the variation of e_1^H.

Concerning the lower limit for electricity production in period 1 in the second line in (9.9) it should be noted that because electricity is non-negative, we have to exclude the possibility of a negative value if the available water is less than the maximal reservoir amount, as is the case for the dry period that will be shown in Figure 9.3 below. In the general case it may happen in many periods that the available water is less than the reservoir limit, because the reservoir limit is without a period subscript and the same for all periods, and this limit will become relatively larger and larger compared with inflows as the period length is decreased. A reservoir limit of 70% of the normal yearly inflow, as is the case for Norway, means that the inflow for an average week is less than 3% of the reservoir capacity, or put it another way: for an average week the reservoir level at the end of the previous week must represent a filling of more than 97% for more than the reservoir content to be available. When the available water in a period falls short of the reservoir limit we cannot have a corner solution of transferring the total amount available to the next period, but must have an interior solution or the corner solution of transferring zero (corresponding to the illustration for the dry period in Figure 9.3 below). When having the maximal transfer from a period to the next as a corner solution we will therefore have the situation that the available water in a period receiving a full reservoir necessarily exceeds the reservoir limit if the realised inflow is positive.

The first-order condition for determining the optimal value of consumption in period 1 for an interior solution is[2]:

$$p_1(e_1^H) - E\{p_2(R_o + w_1 + w_2 - e_1^H)\} = 0 \qquad (9.10)$$

[2] It is assumed that the probability distribution for the inflow is regular in the sense that the last expression in (9.10) is valid.

For values of consumption in the interior of the interval specified in (9.9) the price in period 1 is set equal to the expected price in period 2. The consumption e_1^H in period 1 is in principle determined implicitly from the equation. Specifying a distribution function for the probability of inflows would permit a solution for the consumption in the first period to be found, e.g., by numerical methods.

In practical applications the distribution for the inflow is discretised, using information about past inflows for the time period in question. In Norway there are data going back about 70 years. The frequencies of a sufficiently high number of inflow outcomes (measured as total inflow within suitable intervals) can be calculated as the average numbers for 70 years, including the levels of w_2^{min} and w_2^{max}.[3] The expected price in period 2 can then be expressed as:

$$E\{p_2(e_2^H)\} = \sum_{i=1}^{K} \phi_i p_2(R_o + w_1 + w_i - e_1^H) \tag{9.11}$$

where ϕ_i is the non-negative frequency for the inflow in the interval i and K is the total number of intervals. This expression can be used when solving (9.10) for the production level of period 1, specifying also the demand functions. Knowing the production level in period 1 the transfer of water R_1 to period 2 is readily provided by the water accumulation equation $R_1 = R_o + w_1 - e_1^H$.

A standard property of the demand function is that it is convex, as we have assumed throughout the book. It then follows from *Jensen's inequality* that

$$E\{p_2(R_o + w_1 + w_2 - e_1^H)\} \geq p_2(R_o + w_1 + E\{w_2\} - e_1^H) \tag{9.12}$$

Equality holds if the demand function is linear. The price in period 1 should in general be set *higher* than the price formed for period 2 by using the expected inflow in the demand function for period 2, given the optimal consumption in period 1 and amount of water transferred to period 2 from period 1. Although we cannot, strictly speaking, compare the solution for uncertainty with the deterministic case treated in Chapter 3, using the expected amount of inflow is often used as a benchmark. Convexity of the demand function implies that the possibility of realising low inflows and correspondingly high prices in period 2, results in less consumption in period 1 and a greater transfer of water to period 2 than a naïve prediction of the price in period 2, by applying the expected inflows in period 2 in the

[3] The extreme values of the distribution may be estimated using approaches for extreme-value estimation.

demand function, would yield. The effect of convexity of the demand function is to create a relatively higher increase in price if the realisation of inflow in period 2 should turn out to be low than the relative decrease in price if the realisation turns out to be high. Compared with the benchmark of equal prices (in the case of the reservoir constraint not being binding), the social planner strives to make the prices as equal as possible in the face of uncertainty. In order to correct for the tendency for the *ex post* difference to be higher for less water than plenty of water, production is reduced in period 1.

We have assumed risk neutrality on behalf of the social planner. Uncertainty has an unavoidable social cost. This cost is exposed by the difference between the period prices when we have moved to period 2. Introducing risk aversion would probably reinforce the effect convexity of the demand function gives, since periods with exceptionally high electricity prices are known to cause political stress.

If the risk of extreme events increases in the sense of *mean-preserving spread* (Rothchild and Stiglitz, 1970) it follows directly from that paper that $E\{p_2(R_o + w_1 + w_2 - e_1^H)\}$ increases when the demand function is convex. This implies that the price in period 1 is set higher for increased uncertainty, in the sense of mean preserving spread, about the inflows in period 2.

Corner solutions for consumption in period 1 appear when the condition (9.10) yields values of consumption in period 1 outside the admissible interval. In (9.8) the upper limit for the price in period 1 is calculated for consumption hitting the lower limit. Similarly we get a lower limit for the price in period 1 when hitting the upper limit for consumption in period 1:

$$p_1^{\min} = p_1(R_o + w_1) \tag{9.13}$$

The complete solution of problem (9.9) then follows from the conditions:

$$\begin{aligned}
&p_1(e_1^H) = E\left\{p_2(R_o + w_1 + w_2 - e_1^H)\right\} \\
&\text{for } e_1^H \in \left(\max(0, R_o + w_1 - \bar{R}), R_o + w_1\right) \\
&E\left\{p_2(\bar{R} + w_2)\right\} \geq p_1^{\max} \equiv p_1(R_o + w_1 - \bar{R}) \Rightarrow e_1^H = R_o + w_1 - \bar{R}, \\
&E\left\{p_2(w_2)\right\} \leq p_1^{\min} \equiv p_1(R_o + w_1) \Rightarrow e_1^H = R_o + w_1
\end{aligned} \tag{9.14}$$

The determination of the transfer from period 1 to period 2 follows directly from using the water accumulation equation (9.2).

The optimal choice of consumption in period 1 is illustrated in Figure 9.3. Only the decision to be made in period 1 is shown. The left-hand vertical axis measures the price in period 1 and the horizontal axis measures

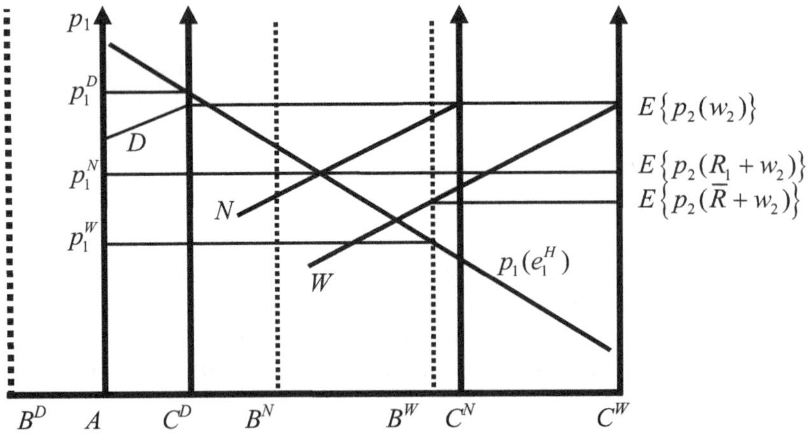

Figure 9.3. Optimal consumption in period 1.

the available water in period 1. Three situations for period 1are portrayed; a dry period having only the amount AC^D at disposal, a normal period having AC^N at disposal, and a wet period having AC^W at disposal. In the dry period less water is at disposal than the capacity of the reservoir. The right-hand axes are erected at the end points of the available water in each of the periods. The same reservoir capacity in each period is indicated by B^DC^D, B^NC^N and B^WC^W respectively, measured from right to left from the end point of available water of the three alternatives. The broken lines are erected from the B-points. The curves labelled D, N and W show how the expected price expressed in (9.14) in period 2 varies with the amount transferred from period 1 to period 2 for the three different situations as to water availability in period 1. The amount transferred varies from the maximal AC^D, B^NC^N and B^WC^W respectively from the intersections of the curves and the vertical lines erected from A, B^N and B^W respectively to zero at the intersections with the vertical lines erected from C^D, C^N and C^W. (The curves end at the respective right-hand axes at a lower value than the choke price for period 1 for the convenience of the illustration, and this does not reflect a general feature.)

When period 1 is a dry period the value of the expected price in period 2 is lower than the minimum equilibrium price in period 1 corresponding to all available water being consumed in period 1. It is then not optimal with any transfer of water to period 2, in accordance with the last condition in (9.14). When period 1 is a normal period then some of the available water in period 1 is transferred to period 2 (but not as much as the maximal reservoir), and this interior solution implies that the price in period 1 is set equal to the expected price in period 2, in accordance with the first condition in (9.14). When period 1 is a wet period, even transferring the whole reservoir to period 2 is not enough to make the price in period 1 as high as the expected price in period 2, in accordance with the second condition in (9.14).

In addition to consider different patterns of inflows in period 1, it may also be of interest to only consider one type of inflow regime in period 1, but to consider different probability distributions for period 2. This will be especially useful for generalising to many periods. Three different distributions for the inflow in period 2 is considered in Figure 9.4 for period 1, termed dry period (*D*), normal period (*N*), and wet period (*W*), respectively. The expectations shown in the figure are made conditional upon the three different distributions for the inflow in period 2 using *D*, *N*, and *W* as symbols. The width of the figure corresponds to the available water for period 1; that is what was in the reservoir at the start of period 1 and the inflow in period 1, both known quantities. *AC* measures the available water, and *BC* shows the size of the reservoir. Demand in period 1 is measured from the left-hand axis. The expected price in period 2 for the three different distributions as function of the amount of water transferred from period 1 to period 2, is measured from the right-hand vertical axis to the left,

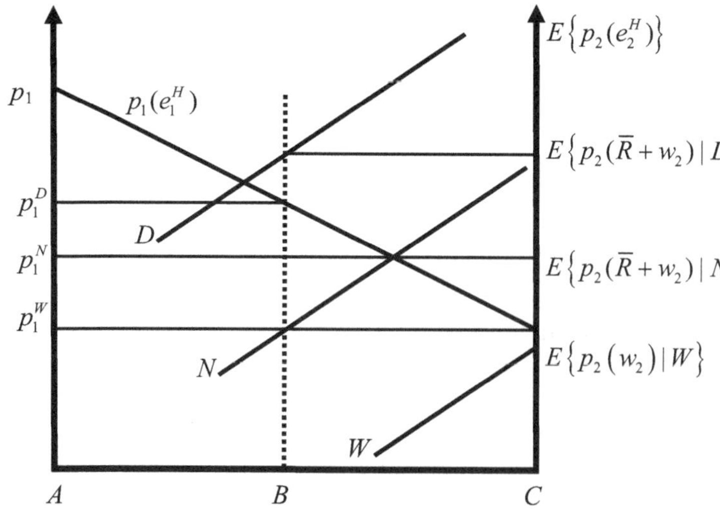

Figure 9.4. Optimal price and consumption in period 1.
Stochastic inflow in period 2 follows three different distributions.

starting with zero water transferred and then decreasing as more and more water is transferred. A dry period in period 2 means that this curve must give higher values for the expected price as a function of transferred water than is the case for both a normal period and a wet period, the latter yielding the lowest-placed curve in the figure. This follows directly from the different expectations about inflows in period 2.

Considering the dry-period scenario first, the intersection of the demand curve for period 1 and the expected period 2 price curve is to the left of the limit of the reservoir. This implies that we have a corner solution for the transfer from period 1 to period 2: expecting a dry period in period 2 the maximal amount $\overline{R}$ is transferred from period 1 to period 2. In order to obtain this transfer without overflow in period 1 the price p_1^D has to be charged. The expected price $E\{p_2(\overline{R}+w_2)|D\}$ in period 2 is higher than this price; $E\{p_2(\overline{R}+w_2)|D\} > p_1^D$, in accordance with the first corner solution in (9.14).

Adopting the normal expectation about the inflow in period 2 takes us to the case of the expected price curve for period 2 labelled N in the figure. The intersection of the demand curve for period 1 and the curve for the expected price in period 2 is within the reservoir capacity. This implies that we have an interior solution for electricity consumption in period 1, yielding the price in period 1 equal to the expected price in period 2 in accordance with (9.14), $E\{p_2(R_1 + w_2)|N\} = p_1^N$.

Expecting period 2 to be wet the bottom curve labelled W for the expected price in period 2 is valid. This curve does not intersect with the demand curve for period 1. The implication is that we have a corner solution with no transfer of water from period 1 to period 2. The price in period 1 is set at p_1^W implying all available water is demanded in period 1. The expected price in period 2 is $E\{p_2(w_2)|W\}$, and we have $E\{p_2(w_2)|W\} < p_1^W$, in accordance with the second corner solution in (9.14).

When time passes and we move on to period 2 the optimal decision for consumption in period 2 is to consume all available water and empty the reservoir. The actual inflow may then not be as expected when the decision for transfer of water from period 1 to period 2 had to be made. The distribution of possible *ex post* outcomes is illustrated in Figure 9.5. The range of actual realisations of the inflow in period 2 is between the minimal inflow and the maximal, generating the gap between actual realised optimal period-2 prices, corresponding to the maximal gap between inflows. The expected price for period 2, standing in period 1, is written $p_1 = E_1\ \{p_2\}$ in the figure. We know from Chapter 3 that periods with scarcity and periods with threat of overflow are price-determining events. In the two-period

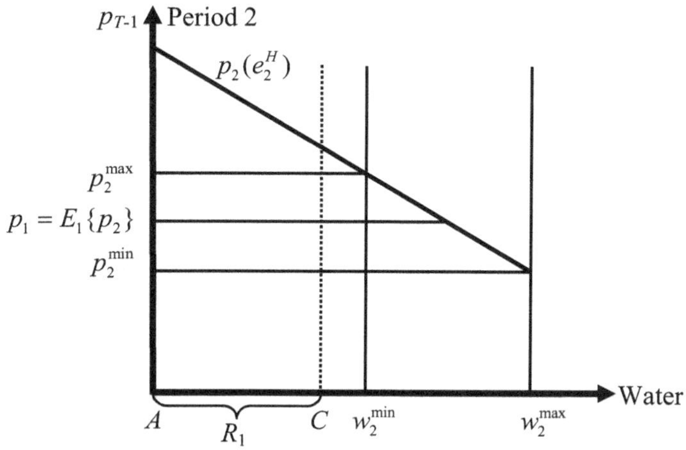

Figure 9.5. Possible price range when in period 2.

model there is scarcity in period 2. This is reflected in the decision rules for consumption in period 1 formulated in (9.14). But as seen from Figure 9.5 the actual realisation of the level of scarcity in period 2 can generate a wide distribution of the realised period 2 price. An important conclusion is then that deviations between expectations and realisations of inflows may generate differences in price over time. Uncertainty contributes independently to price variation. In the deterministic case in Chapter 3 the prices in period 1 and period 2 should be equal in the case of the reservoir constraint not becoming binding in period 1. Due to uncertainty the prices will now as a general rule differ. This reflects the fact that a decision about use of water today must be based on expected inflows tomorrow, and that it would be arbitrary that the expectation is realised exactly.

Generalisation to *T* periods

Some main features of the situation with uncertain inflows were revealed using just two periods, but not all. We will need to consider at least three periods to see some complications working out an optimal solution standing in the starting period, and then we may as well try to characterise the solution for *T* periods. We will not try to give a complete account of how to establish a solution in the general case, but indicate the main steps. The purpose is just to establish that uncertainty can generate price variation that would not be there in the deterministic case.

Following the principle of backwards induction to ensure a consistent optimal plan in a dynamic world, we start with the terminal period T. Due to the terminal condition that the reservoir level at the end of period T is free, the reservoir is emptied under the assumption of demand not being satiated, i.e., $R_T = 0$. When we are in period T the inflow is known, and R_{T-1} is known from the past, so we simply get:

$$e_T^H = R_{T-1} + w_T\,,\ p_T = p_T(R_{T-1} + w_T) > 0\,, \tag{9.15}$$

as discussed in Chapter 3. The amount transferred from period $T - 1$ is in the interval $[0, \overline{R}]$. If the amount transferred is zero, then it is possible that the realised inflow in period T is zero if this value is permitted by the probability distribution. This occurrence implies that the choke price, assumed finite, is realised.

Moving to period $T - 1$ the inflow in period T is then stochastic. The price in period T will therefore be an expected price. The solution for the price- and production level in period $T - 1$ follows directly from adapting (9.14), using the period index $T - 1$ instead of 1:

$$\begin{aligned}
&p_{T-1}(e_{T-1}^H) = E\left\{p_T(e_T^H)\right\} = E\left\{p_T(R_{T-1} + w_T)\right\} \\
&\quad \text{for } e_{T-1}^H \in \left(\max(0, R_{T-2} + w_{T-1} - \overline{R}), R_{T-2} + w_{T-1}\right) \\
&p_{T-1}(e_{T-1}^H) \geq E\left\{p_T(\overline{R} + w_T)\right\} \\
&\quad \text{for } e_{T-1}^H = R_{T-2} + w_{T-1} - \overline{R} > 0 (R_{T-1} = \overline{R}) \\
&p_{T-1}(e_{T-1}^H) \leq E\left\{p_T(w_T)\right\} \\
&\quad \text{for } e_{T-1}^H = R_{T-2} + w_{T-1}\ (R_{T-1} = 0)
\end{aligned} \tag{9.16}$$

The production level in period $T - 1$ is found implicitly by substituting for R_{T-1}, using the water accumulation equation, in the second equation in (9.16) in order to bring in e_{T-1}^H as a variable:

$$p_{T-1}(e_{T-1}^H) = E\left\{p_T(R_{T-1} + w_T)\right\} = E\left\{p_T(R_{T-2} + w_{T-1} - e_{T-1}^H + w_T)\right\} \tag{9.17}$$

The production level will be a function of the non-stochastic variables R_{T-2}, assumed known from the past, w_{T-1} known in the current period, and the stochastic variable w_T. Having the solution for the production level the amount transferred to the next period is determined by using the water accumulation equation again:

$$R_{T-1} = R_{T-2} + w_{T-1} - e_{T-1}^{H*} \tag{9.18}$$

where e_{T-1}^{H*} is the solution for electricity production from (9.17). Notice that both the price in period $T - 1$ and the production level are functions of the amount of water handed over from period $T - 2$. The two corner solutions for water transferred from period $T - 1$ to T under our assumptions yield the optimal level of production in period $T - 1$ directly from (9.16), and the levels are also functions of the water transferred from period $T - 2$.

If we focus on the price levels in period $T - 1$ and T and remember the one-to-one correspondence between water values and social prices, the first expectation expression in (9.17) gives us the price in period $T - 1$ equal to the expected water value of period T, conditional on the transfer of water from period $T - 1$ to T:

$$p_{T-1} = E\left\{p_T \middle| R_{T-1}\right\} \tag{9.19}$$

Going through admissible values for the amount of transferred water (including corner solution values) the right-hand expression gives us all the possible expected values of the water value in period T, calibrated for a given value of R_{T-2}. Such a function may be termed the *expected water value table* corresponding to a concept used in the literature (Hveding, 1967, 1968). The information given by such a "table" may be utilised determining the actual quantities and prices as time evolves from the start of the planning period. This table was actually used in the two-period case shown in Figures 9.3 and 9.4.We will return to this point below.

Moving to period $T - 2$, following the same type of substitutions as above, we have

$$\begin{aligned} &p_{T-2}(e_{T-2}^{H}) = E\left\{p_{T-1}(e_{T-1}^{H})\right\} = E\left\{p_{T-1}(R_{T-2} - R_{T-1} + w_{T-1})\right\} = \\ &E\left\{p_{T-1}(R_{T-3} + w_{T-2} - e_{T-2}^{H} - R_{T-1} + w_{T-1})\right\} \end{aligned} \tag{9.20}$$

This equation can be solved for the production level of period $T - 2$, given a value of the transfer from period $T - 3$ to period $T - 2$. A new feature seen for period $T - 2$ is that the amount of water transferred from period $T - 1$ to T is also appearing. For period $T - 1$ we knew that $R_T = 0$.The value for R_{T-1} is determined in the previous round for period $T - 1$ together with the production level for that period [see (9.18)] as a function of R_{T-2}, w_{T-1} and the stochastic variable w_T. The corner solutions follow as in (9.16), updating the time index, using the two extreme values for the amount transferred to period $T - 1$. The solution for the current period $T - 2$ involves the solution for the previous period T. We can also say that the expectation of the solution for period T is contained in the expected water value.

When forming the expected water value table for use in period $T-2$ we now have the new feature that the amount of water transferred from period $T-1$ to T enters the expression for the amount produced in period $T-2$. The expected price in period $T-1$ is then conditional both on the transfer of water from period $T-2$ to $T-1$, and the transfer of water from previous period $T-1$ to T. The transfer of water from period $T-1$ to T is a stochastic variable. It is natural to express the water-value table updating (9.19) one period, where the expectation of R_{T-1} is now included in the expectation operation:

$$p_{T-2} = E\{p_{T-1} | R_{T-2}\} = E\{p_{T-1} | R_{T-3} + w_{T-2} - e_{T-2}^H\} \tag{9.21}$$

The expected water-value table for period $T-1$ is now calibrated for a value of the water transferred from period $T-3$.

Following the general substitution principle the conditions determining the price and quantities for period t are:

$$\begin{aligned}
&p_t(e_t^H) = E\{p_{t+1}(e_{t+1}^H)\} = E\{p_{t+1}(R_t - R_{t+1} + w_{t+1})\} \\
&\quad \text{for } e_t^H \in \left(\max(0, R_{t-1} + w_t - \bar{R}), R_{t-1} + w_t\right) \\
&p_t(e_t^H) \geq E\{p_{t+1}(\bar{R} - R_{t+1} + w_{t+1})\} \\
&\quad \text{for } e_t^H = R_{t-1} + w_t - \bar{R} > 0 \ (R_t = \bar{R}) \\
&p_t(e_t^H) \leq E\{p_{t+1}(w_{t+1} - R_{t+1})\} \\
&\quad \text{for } e_t^H = R_{t-1} + w_t \ (R_t = 0)
\end{aligned} \tag{9.22}$$

The output level e_t^H is implicitly determined for the reservoir level in period t substituting in the second expression in the first condition in (9.21) for the transfer of water from period t to $t+1$:

$$\begin{aligned}
p_t(e_t^H) &= E\{p_{t+1}(R_t - R_{t+1} + w_{t+1})\} \\
&= E\{p_{t+1}(R_{t-1} - R_{t+1} + w_t + w_{t+1} - e_t^H)\}
\end{aligned} \tag{9.23}$$

Here the value of R_{t+1} is determined in the previous round in period $t+1$ as a function of R_{t-1}, the stochastic variables w_{t+1} and w_{t+2}, and R_{t+1}, also being a stochastic variable.

It may be informative to carry out substitutions in (9.23) to bring out the point that the solution for the production level for period t depends on the solutions for all later-period quantities. Using the water-accumulation equation yields:

$$p_t(e_t^H) = E\left\{ p_{t+1}(R_{t-1} + \sum_{i=t}^{T} w_i - e_t^H - \sum_{i=t+2}^{T} e_i^H) \right\} \; (t = 1,..,T-2) \tag{9.24}$$

The expected water-value table relevant for period t is, generalising (9.21): $E\{p_{t+1}|R_t\} = E\{p_{t+1}|R_{t-1} + w_t - e_t^H\}$. The expectation-operation is carried out with respect to all the stochastic variables appearing in (9.24), thus showing the dependence on the earlier (going backwards) solutions for the production levels. The expected water-value table used in period t is calibrated on the transfer of water from period $t-1$ to t.

We have that the arguments in the p_{t+1}-function involve the water transferred from period $t-1$ to t, the total inflows from t to T and the water use from t to T, excluding what is used in period $t+1$, summing up to the use of water in period $t+1$.

In order to solve for prices and quantities, standing in the planning period $t = 1$, we have to find both the expected value for the transfer of water to the period we are considering and the expected value of the water transferred from the period after the period we are considering. The latter involve the solutions for all production levels from two periods after the one we are considering right back to the terminal period. All the demand functions are then involved also. This is the challenge to the algorithm that has to be set up to solve the problem numerically.

Going backwards to the start $t = 1$ of the planning period we get a solution for the use of water in period 1 and the transfer to period 2 as functions of R_o and w_1, both of which are known in period 1. But as noticed above we need to use the expected price for period 2. The expectation involves the transfer of water from period 2 to period 3 being solved backwards for the terminal period T and right to period 3 involving substitution using the dynamic water accumulation equation as shown in Eq. (9.24). It should be noted that the solution for a period depends on *future* stochastic variables, thus the solutions for remaining periods will not be revised as time passes (assuming stochastic independence between periods). No new information concerning the solutions for the remaining periods is revealed by the passing of time.

Moving forward in real time it will be arbitrary that the expected price formed at the start in period 1 is realised in later periods, i.e., we will generally have $p_t \neq E\{p_{t+1}|R_t\}$ when we have moved to period t. The actual realisations of the inflows and the deviations from the whole expected time path will generate fluctuations in the price level. The mechanism can be illustrated in Figure 9.6. In period t the available water is AC ($= R_{t-1} + w_t$). The size of the reservoir is BC. The expected water value table to be used

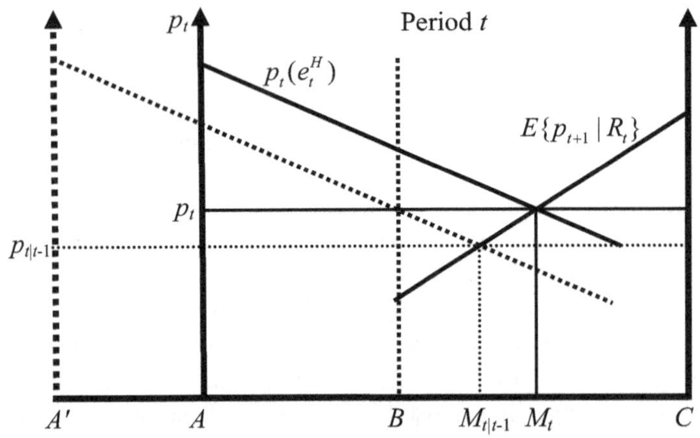

Figure 9.6. The actual adjustment when real time has moved to period t.

in period t is $E\{p_{t+1}|R_t\}$ and the price for period t is set following the intersection of the demand curve for period t and the curve representing the variation in expected water values with the amount of water transferred from period t to period $t + 1$. This curve has been calibrated according to the historic value for R_{t-1} that may deviate from the expected value in the optimal plan. When looking forward, being in period $t - 1$, a greater inflow (or a greater transfer from $t - 1$ to t) represented by $A'C$ was expected. (Notice that the point C is kept fixed, it is the point A that is moved.) The demand curve was expected to be anchored at the dotted wall from A', and the expected price for period t formed at period $t - 1$ is indicated in the figure as $p_{t|t-1}$. The production in period t was expected to be $A'M_{t|t-1}$ and the transfer to period t + 1expected to be $M_{t|t-1}C$. The actual price for period t set in the same period is higher than the expectation formed in the previous period, resulting in a lower production in period t, but also a lower transfer, M_tC, to period $t + 1$. With a sufficient deficit in available water we may get a corner solution with zero transfer of water to the next period $t + 1$.

With more water available in period t than expected the effects will be opposite (the interpretation of AC and $A'C$ in the figure can be switched). The available water in period t may become so abundant as to result in a corner solution of transferring the maximal amount of the whole reservoir to period $t + 1$. It should be noticed that these results are general because the expected water value curve remains fixed, anchored at the wall erected from C, and is by construction in the same position independent of the realisation of available water.

Standing in the starting period 1 and looking forward at the expected price path this path reflect corner solutions according to (9.22) and in general may mimic the price structure discussed in Chapter 3. However, as we move forward in real time the corner solutions may not appear in the expected periods. The same mechanism as discussed above may also lead to deviations between the expected corner solution periods and the actual corner solutions taking place. We see from Figure 9.6 that an expected episode with threat of overflow may be postponed if the realised available water in the expected period with a full reservoir is less than expected. The actual period with threat of overflow may come earlier if more water than expected is available in periods leading up to the expected period with a full reservoir, since more water will then be transferred to the next period.

Hydro and thermal

Keeping all variables involved with thermal production deterministic, it may still be of interest to study whether there are any consequences for the combined utilisation of hydro and thermal when assuming stochastic inflows. Only the simplest situation of two periods is considered. Thermal capacity is represented by an aggregated convex cost function in total output based on merit-order ranking, based on marginal costs of individual generators, as explained in Chapter 5. The thermal cost function in period 2 is assumed to be equal to the function in period 1 and known with certainty. A generalisation would be to consider future fuel prices to be stochastic. The problem with hydro and thermal capacities for two periods can be set up as follows when inflow is stochastic in the second period:

$$
\begin{aligned}
&\max \left[\int_{z=0}^{x_1} p_1(z)\,dz - c(e_1^{Th}) + E\left\{ \int_{z=0}^{x_2} p_2(z)\,dz - c(e_2^{Th}) \right\} \right], \\
&x_t = e_t^H + e_t^{Th}, t = 1,2 \\
&e_1^H \in \left[\max(0, R_o + w_1 - \overline{R}), R_o + w_1 \right] \\
&e_2^H \in \left[w_2, \overline{R} + w_2 \right] \\
&e_t^{Th} \in \left[0, \overline{e}^{Th} \right], t = 1,2
\end{aligned}
\tag{9.25}
$$

Substituting for total electricity production (= consumption) from the energy balance, and using the water accumulation equation for period 2 we get problem (9.9) expanded by adding thermal capacity:

$$\max_{e_1^H, e_1^{Th}} \left[\int_{z=0}^{e_1^H + e_1^{Th}} p_1(z)dz - c(e_1^{Th}) + E\left\{ \int_{z=0}^{R_o + w_1 + w_2 - e_1^H + e_2^{Th}} p_2(z)dz - c(e_2^{Th}) \right\} \right]$$
$$e_1^H \in \left[\max(0, R_o + w_1 - \overline{R}), R_o + w_1 \right]$$
$$e_t^{Th} \in \left[0, \overline{e}^{Th} \right], t = 1,2 \tag{9.26}$$

The first-order conditions for interior solutions are:

$$p_1(e_1^H + e_1^{Th}) - E\left\{p_2(R_o + w_1 + w_2 - e_1^H + e_2^{Th})\right\} = 0$$
$$p_1(e_1^H + e_1^{Th}) - c'(e_1^{Th}) = 0 \tag{9.27}$$
$$E\left\{p_2(R_o + w_1 + w_2 - e_1^H + e_2^{Th})\right\} - E\left\{c'(e_2^{Th})\right\} = 0$$

The price in period 1 is set equal to the expected price in period 2. Furthermore, the marginal costs in period 1 is equal to the expected marginal cost in period 2, and both are set equal to the expected price in period 2. The use of thermal capacity in period 2 becomes a stochastic variable because the inflow in period 2 is stochastic. In principle we have three equations to determine the three endogenous variables. The use of thermal capacity in period 1 can be solved as a function of the hydro production in period 1 from the second condition in (9.27).

Using the water accumulation equation the first expression in the last equation in (9.27) can be written $E\{p_2(R_1 + w_2 + e_2^{Th})\}$. Increasing R_1 assuming the other variables fixed will reduce the expected price. However, the thermal capacity in period 2 is endogenous and a reduction in the expected price will imply an expected reduction in the thermal capacity, thus counteracting the reduction in expected price. The last condition in (9.27) can be written:

$$E\left\{p_2(R_1 + w_2 + e_2^{Th})\right\} = E\left\{c'(e_2^{Th})\right\} \tag{9.28}$$

For a given R_1 this equation yields a relationship between w_2 and e_2^{Th}. When writing the expectations expression on the left-hand side above as $E\{p_2|R_1\}$ we assume that both the expectation of the inflow and the thermal capacity in period 2, conditional on the choice of reservoir level at the end of period 1, is covered.

Figure 9.7 provides an illustration of the solution to the choices in period 1of hydro and thermal production, and the amount of water to be transferred to period 2. The known amount of water is AC, and the reser-

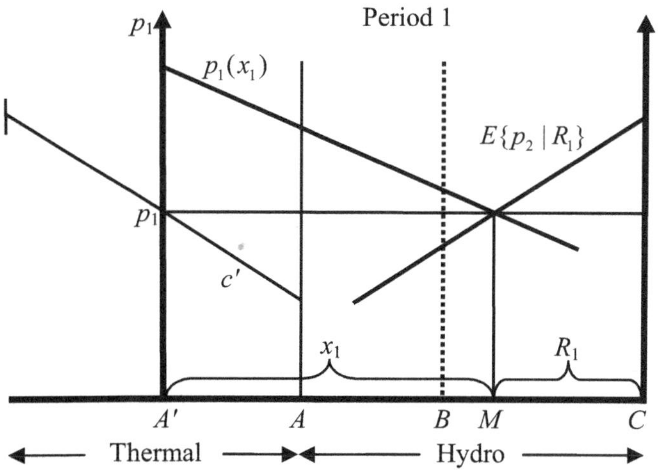

Figure 9.7. Hydro and thermal.

voir capacity is BC. The marginal cost function for the thermal capacity is shown by the curve marked c', anchored at the hydro wall up from A, and measured from right to left. The thermal wall is erected from A'. The placement of the wall is endogenous. The demand function is anchored on the thermal wall. The expectation-function is anchored on the hydro wall on the right-hand side of the figure for a zero value of the transfer from period 1 to period 2, and intersects the reservoir capacity curve represented by the broken line from B at maximal transfer. Due to the interaction effect with thermal capacity, the slope of the expectation curve should be less steep than the slope of the comparable curve in the case of hydropower only in the previous section, assuming that total electricity amounts involved are the same. Equilibrium is found as the intersection of the demand curve and the expectation curve. The price for period 1 will determine the amount of thermal capacity, $A'A$, taken into use as the intersection of the equilibrium price in the figure and the marginal cost curve. The point M, corresponding to the intersection of the demand- and expectations curve, shows the allocation of water on consumption in period 1, AM, and transfer to period 2, MC.

There may be corner solutions for the thermal capacity in period 1 if the expected price becomes lower than $c'(0)$ or higher than $c'(\overline{e}^{Th})$. The corner solutions for hydro corresponds to the solutions in (9.14), but thermal capacity has to be introduced, with its upper and lower constraint.

When moving to period 2 the water inflow becomes known, and the use of thermal in period 2 is decided as in Chapter 5 with equalisation of mar-

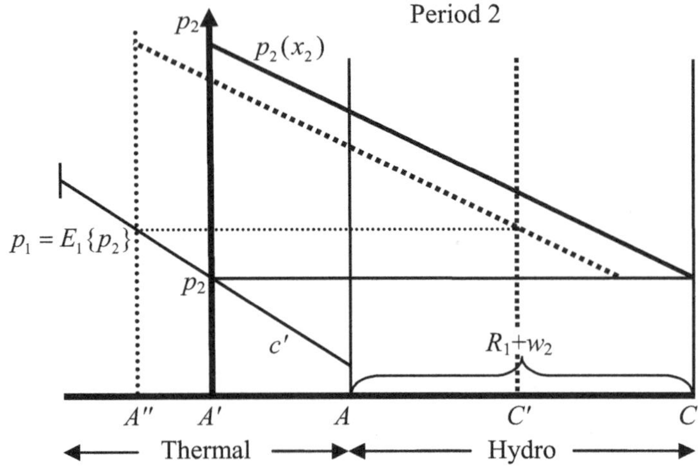

Figure 9.8. Hydro and thermal in period 2.

ginal cost and price in period 2. The situation is illustrated in Figure 9.8. The available water, AC (= R_1 + w_2), is utilised together with the thermal capacity such that the amount $A'A$ is used, according to equalisation of price and marginal cost. If in period 1 the realised inflow becomes greater than expected, the expected price in period 2, indicated as the horizontal dotted line $p_1 = E_1(p_2)$ in the figure, should be higher than the realised price. Expected available water in period 2 was AC' and expected use of thermal capacity $A''A$, as indicated by the dotted lines. The expected placement of the demand curve is correspondingly shown by the broken line as a shift to the left of the demand curve. The opposite movement in utilisation of thermal capacity dampens the deviation of price from the expected. For the same amounts of electricity in the two periods the possible price differences in the case of hydropower only, shown in Figure 9.5, should be greater than in the case with thermal, as can be indicated elaborating the limits of the inflow distribution in Figure 9.8.

Monopoly revisited

We want to investigate whether uncertainty about future inflows will change the way a monopolist finds it profitable to shift the water from relatively inelastic demand periods to relatively elastic periods that we investigated in Chapter 8 under full certainty. The model is as simple as possible with two periods and total amount of water as the constraint, following model (8.1) in Chapter 8. The inflow is known in period 1, and we investigate the case that the upper reservoir constraint will never be binding. The inflow in the second period is stochastic seen from period 1. The total available water in period 2, $W - e_1^H$, is therefore stochastic. The profit-maximising problem of the monopolist is:

$$\max\left[p_1(e_1^H)\cdot e_1^H + E\left\{p_2(e_2^H)\cdot e_2^H\right\}\right]$$
$$\text{subject to} \tag{9.29}$$
$$\sum_{t=1}^{2} e_t^H \leq W, W \text{ stochastic}$$

An additional formal requirement in the model is that all variables are non-negative. Since the water-accumulation equation is not explicitly modelled, any spilling of water will appear as being done in the second period. Our model formulation is *as if* all water is also available in period 1. Water accumulation has to be shown explicitly to identify spilling in period 1.

Inserting the total available water in the expression for expected profit in period 2 the maximisation problem becomes:

$$\max\left[p_1(e_1^H)\cdot e_1^H + E\left\{p_2(W - e_1^H)\cdot(W - e_1^H)\right\}\right] \tag{9.30}$$

The necessary first-order condition is:

$$\begin{aligned} & p_1(e_1^H) + p_1'(e_1^H)\cdot e_1^H \\ & \quad - E\left\{p_2(W - e_1^H) + p_2'(W - e_1^H)\cdot(W - e_1^H)\right\} = \\ & p_1(e_1^H)(1 + \breve{\eta}_1(e_1^H)) \\ & \quad - E\left\{p_2(W - e_1^H)(1 + \breve{\eta}_2(W - e_1^H))\right\} = 0 \end{aligned} \tag{9.31}$$

In the second equality the (negative) price flexibilities are introduced, defined as $\breve{\eta}_t = p_t' e_t^H / p_t \ (t = 1,2)$. As is standard for the monopoly problem to make economic sense, we must have $e_t^H, p_t(e_t^H) > 0$, $(1 + \breve{\eta}_t) \geq 0 (t = 1,2)$, as discussed in Chapter 8. The first-order condition requires that the marginal revenue, or flexibility-corrected price, in period 1 must be equal to the expected marginal revenue (flexibility-corrected price) in period 2. Knowing the probability distribution for W the solution for production in the first period can be found implicitly from (9.31).

Shifting of water between periods now takes place based on comparing a known flexibility with an expected one. As in the deterministic case covered in Chapter 8, compared with the social allocation of water the monopolist will use more water in a relatively elastic demand period and less in a period with relatively inelastic demand. But the economic success of the shifting policy is only seen *ex post* in period 2 when a value of the available water is realised.

Spilling will be expected if the optimality condition in (9.31) turns out to require the marginal revenues to be equal to zero:

$$p_1(e_1^H)(1+\breve{\eta}_1(e_1^H)) = E\left\{p_2(W-e_1^H)(1+\breve{\eta}_2(W-e_1^H))\right\} = 0 \qquad (9.32)$$

This condition implies that the demand flexibility in period 1 has the absolute value of 1, and that this also holds in an expected sense for period 2. When moving to period 2 the amount of water available becomes known, and spilling may or may not be optimal depending on the realisation of the inflow of water.

In the social planning case with uncertainty we used Jensen's inequality to show that the expected price in period 2 was higher than inserting the expected consumption for period 2 in the demand function [see (9.12)] if the demand function was convex. Furthermore, the Rothchild and Stiglitz (1970) result for mean-preserving spread was invoked to show that less water is used in period 1 the higher the probability of extreme events if the demand function is convex. Comparing the case of uncertainty with the deterministic case for the monopolist, we have that the same holds provided that the marginal revenue function is convex. But this property does not follow from convexity of the demand function. We see from (9.31) that the *third* derivative of the demand function is involved in determining whether the marginal revenue function is convex.[4] Assuming convexity, the higher the probability of extreme events the higher value of the expected flexibility-corrected price in period 2, and the more water should then be used in period 1. This leads to two interesting observations. First, since the flexibilities are functions of the amount of electricity involved, there may be a reversal of the period that has the highest (or lowest) demand flexibility. Second, less water will be used in period 1, irrespective of whether this period has the relatively most or least elastic demand, the higher the uncertainty in a mean-preserving spread sense. The economic rationale for this is that convexity of the marginal revenue function means that the greatest difference between the marginal revenues will occur if less water is realised in period 2. The loss for the monopolist of not "hitting the target" of

[4] Consequences of convex marginal revenue function are investigated in Hansen (2007) using explicit parameterisation of variance. It is pointed out that a concave marginal revenue function leads to the opposite conclusion. This is, of course, also valid for the present analysis.

equality of the marginal revenues is greater the greater the differences between the marginal revenues turn out to be *ex post*.

As regards exercising market power an interesting question is if uncertainty contributes to market power being more or less felt, i.e., does the monopoly solution under uncertainty deviate more from the social planning solution under uncertainty than is the case under no uncertainty? It is not easy to give a qualitative answer. We saw that in the case of uncertainty both in the social planning case and for a monopolist the first period production will be reduced, compared with a situation in which the expected available water is inserted in the demand function. The monopolist will still practice shifting of water. In the second period the best a monopolist can do is to either process all water or spill some water. In the case of no spilling the monopolist will then charge exactly the same price as the social planner, having the same amount of water at his disposal in period 2. Such a situation then seems to imply that uncertainty reduces the scope for market power. To determine whether the welfare loss measured in consumer surplus terms is smaller or greater for the two periods taken together seems to require empirical information about demand functions and the probability function for inflows.

Concluding comments

The presence of uncertainty provides the final reasons for price variation of electricity over time in a hydropower system. In addition to emptying the reservoir and entering a situation with threat of overflow, uncertainty about future inflows will independently create price variations in the social planning solution. Although the problems we set up were quite simple, we saw that to obtain solutions may be a complex task, and has to be done numerically for real-life applications.

One simplification was to specify only one hydropower plant with a single reservoir of a limited size. Extending the uncertainty analysis to multiple plants with one reservoir each, and introducing constraints on the upper production (or power) capacities, and environmental constraints as given in Table 3.1 in Chapter 3, will complicate the analysis considerably. We saw in Chapter 4 that although the social prices are the same for all plants for each period the manoeuvring to avoid overflow is an individual plant task and will now involve the plant-specific uncertainty about inflows. The individual manoeuvring plans must be based on expectations about the future inflows and the social prices, but moving forward in real time not only creates a deviation between the real time price and the expected one, but also implies that each individual plan based on expectations will be subject to adjustments as time evolves. The individual changes then give feedback to the actual price formation within the social planning context.

With uncertainty it would be expected that some overflow would occur. Manoeuvring such that overflow never occurs has a cost that must be weighed against the loss of water when overflow happens. Naturally, *ex ante* the probability of overflow must come into consideration. Morlat (1964) formulated the planning problem under uncertainty analogously to the Hveding conjecture in Chapter 4 about manoeuvring of individual plants that may be termed Morlat's conjecture:

> **Morlat's conjecture:** *Individual reservoirs should be manoeuvred in such a way that the probability of overflow is the same for all reservoirs* (Morlat, 1964, p. 172).

Morlat did not address the situation of emptying the reservoirs, but since the situation is symmetric, it is tempting to suggest that the continuation of Morlat's conjecture would be to state that the manoeuvrings of the reservoirs should also lead to the reservoirs having the same probability of being emptied. However, we will leave this complicated topic here and not attempt to develop a formal analysis.

Chapter 10. Summary and Conclusions

Main drivers of price change

The key theme of the book has been what causes electricity prices to change over time in a hydropower-dominated electricity sector. Regarding water as a limited natural resource, the conclusion for the price structure over time, established in Chapter 2, was that the price should be the same for all periods, in accordance with asset pricing arbitrage a la Hotelling. (Introducing discounting, as would be appropriate for a longer horizon than for the hydropower management problem, would bring in the discount rate in the usual way as Hotelling's rule is expressed, as the growth rate for the electricity price. However, this long-run change is not the type of price change we are talking about here for electricity.) The, maybe surprising, finding in Chapter 2 is that variation in demand over time should not influence the price. The price in, e.g., low-demand summer periods should be the same as the price in high-demand winter periods.

The assumption driving the result about constant price over time, although demand may fluctuate both over the day and over seasons, was that the reservoir limits would never be binding. But in the real world it is generally too costly to have this required reservoir capacity in a hydropower system. Basic events leading to price changes are therefore that a reservoir becomes full or that it is emptied. The way these events may lead to fluctuating prices are extensively analysed in Chapter 3, assuming full certainty and knowledge about future inflows and demand.

However, the number of price changes seemed still to be much less than observed. Introducing restricted generation capacity yielded less manoeuvrability and a break between social price and water value for periods with binding constraint. Together with a reservoir constraint this lead to increased variability when both types of constraints become active. Facing restricted production capacity forced a pattern of use of water that avoided overflow of the reservoirs, thereby reducing the price in periods with non-binding production constraints.

Not all hydropower resources are regulated using reservoirs. Run-of-the-river power may cause extra price variability if reservoirs cannot fully absorb the fluctuation in power availability. The same is the case with power from windmills. Several countries have plans for significant expansion of this source of electricity leading to increased price variability due to the interplay with variable wind and the necessary backup capacity of electricity generation.

A hydropower system usually consists of many power stations and reservoirs with different characteristics as to inflows and relative reservoir capacity. One would believe that such inhomogeneity could add to price variability, but in Chapter 4 the remarkable conjecture of Hveding is established telling us that aggregating individual generation capacities to just one plant and reservoirs to one reservoir is appropriate, provided only individual reservoir constraints are specified. However, the conjecture does not hold introducing individual production constraints. Manoeuvring of individual generators in order to avoid overflow may then influence prices. There is again a divergence between social prices and water values, and prices may also shift due to demand effects. This may also be the case when considering hydrological couplings between plants. Additional constraints with price impacts, especially relevant for short time periods, are environmental constraints regulating water flows and changes in them.

The role of interplay between hydropower capacity and other types of electricity-generating capacity, like thermal capacity, for prices is treated in Chapter 5. The statement that marginal cost of thermal capacity is determining prices in a mixed hydro- and thermal system is often heard. However, in the model analysis the prices are equilibrium prices and cannot be attributed to a specific technology. Hydropower will not be used in periods when the water value is higher than marginal cost of thermal capacity, while thermal capacity will not be used in periods when the water value is lower than even the marginal cost at zero thermal output. When both technologies are in use water value is equal to marginal thermal cost (with an addition of a shadow price on thermal capacity if the latter is exhausted). Having thermal capacity may cause less price variation than in a pure hydro system.

Trade in electricity across national borders is increasingly taking place in Europe. In Chapter 6 the consequences of trade for the price structure in a country were analysed. Taking one country as a point of departure and regarding trade prices as exogenously given, resulted in the trade prices being adopted fully as the prices of the country, when no constraints on interconnector capacity is assumed. A limit on the reservoir capacity did not change the adoption of the trade prices as home prices, just limited the profitability of trade. But restrictions on interconnector capacity lead to price variability within the range of trade prices when interconnector capacity is constrained.

The introduction of a network serving the transmission of electricity from generators to consumers has an impact on the profile of utilisation of stored water and then on the prices. Proper modelling to reflect physical and engineering realities of electricity flowing through networks is a challenging task, and outside the scope of the book. The necessary spatial element of a network was captured by specifying consumption- and production nodes, and implicitly having lines connecting these nodes in a general meshed network, but without modelling loop-flows. Loss was expressed for each of the lines as a function of the flow on the line, and the flow was expressed as a function of all injections at generating nodes and all withdrawals at consumption nodes. This modelling opened up for pervasive network externalities of a change in the spatial configuration of demand over time to influence, in principle, all losses along lines and the spatial distribution of generation. Congestion was modelled as upper constraints on the flow on lines, but without including loop-flow effects. The conclusion from the literature, that spatial node pricing is necessary for an optimal solution emerged. Implementing social spatial pricing not only necessarily led to price variation within a period, but also to impacts on the pattern over time on utilisation of water, generating further price changes.

It is basic knowledge that use of market power can increase prices. In the special case of hydropower monopoly prices will vary between periods due to differences in elasticity of demand, as studied in Chapter 8. The same amount of electricity may be produced within the horizon if spilling is not optimal, but the water use will typically be shifted from periods with relatively inelastic demand to periods with relative elastic demand, thus increasing the variability of prices. Spilling may lead to increased prices in all periods, but is easy to detect and to be prohibited by a regulator.

Uncertainty is a fundamental aspect of hydropower due to the stochastic nature of inflows. In Norway the variability of inflows on a yearly basis may be ± 25 TWh around an average production of 119 TWh, corresponding to a 90 percent confidence interval for inflows. A dry year thus constitutes quite a stress on the system. In Chapter 9 the qualitative impact on electricity prices of dealing with uncertainty was studied within a highly simplified framework. The key decision rule of the social planner facing uncertainty is basing the decision on use of stored water in the current period on the expected price in the next period. The expectation is based on how much water will be transferred to the next period from the current. Expected water values in the next period are formed as a function of the level of the transfer. Such a table is used when deciding production and amount of transfer to the next period in the current period. When moving forward in time expectations will in general not be realised. The optimal reaction to such events is to reduce production in periods with less inflow than expected previously, and increase production in periods that turn out to have more inflow than expected. Thus, the existence of uncertainty leads to a fluctuation in prices unrelated to reservoir constraints and volume of demand. This impact on prices will happen also when adding thermal capacity, although price changes may be dampened.

Competitive electricity markets

In Chapter 4 we investigated the consequence for social planning of many hydropower producers, and found, considering reservoir constraints only, that the system could be treated as one aggregate unit (Hveding's conjecture). We now assume that we are studying one among several suppliers selling electricity in a spot market for every period. There is no uncertainty, so the period price p_t is known. Given the capacity of each producer and the size of his reservoir he will in the situation of no (active) constraint on his reservoir obviously choose to deliver all his electricity in the period with the highest price in order to maximise profits. Therefore, in order to have positive total supply in all periods, a necessary condition is that prices must be equal for all periods in market equilibrium. The allocation over periods is then completely demand driven, and since producers are indifferent about when to produce some additional rule has to be introduced in order to distribute supply according to demand in each period.

In the more realistic case of reservoir and production constraints the situation becomes more complex. Adapting model (4.8) for one producer the constraint set is the same as in the social planning case except that the energy balance does not enter the problem for a single producer. The profit maximisation problem of a producer j is:

$$\begin{aligned}
&\max \sum_{t=1}^{T} p_t e_{jt}^H \\
&\text{subject to} \\
&R_{jt} \le R_{j,t-1} + w_{jt} - e_{jt}^H \\
&R_{jt} \le \overline{R}_j \\
&e_{jt}^H \le \overline{e}_j^H \\
&R_{jt}, e_{jt}^H \ge 0 , \quad t = 1,..,T
\end{aligned} \tag{10.1}$$

$$T, p_t, w_{jt}, R_{jo}, \bar{R}_j, \bar{e}_j^H \text{ given, } t = 1,..,T,$$
$$R_{jT} \text{ free}$$

The electricity production each period and the reservoir filling are the decision variables for the producer. Comparing the social planning problem (4.8) and the profit maximising problem (10.1) of a producer we note that the objective functions are different, and that the market balance equation is dropped from the constraint set.

The Lagrangian for the problem is:

$$\begin{aligned} L = &\sum_{t=1}^{T} p_t e_{jt}^H \\ &-\sum_{t=1}^{T} \lambda_{jt}(R_{jt} - R_{j,t-1} - w_{jt} + e_{jt}^H) \\ &-\sum_{t=1}^{T} \gamma_{jt}(R_{jt} - \bar{R}_j) \\ &-\sum_{t=1}^{T} \rho_{jt}(e_{jt}^H - \bar{e}_j^H) \end{aligned} \tag{10.2}$$

For notational ease we have used the same symbols for shadow prices as in the social planning case with a single producer. The shadow prices are plant specific. The necessary conditions are:

$$\begin{aligned} &\frac{\partial L}{\partial e_{jt}^H} = p_t - \lambda_{jt} - \rho_{jt} \le 0 \; (= 0 \text{ for } e_{jt}^H > 0) \\ &\frac{\partial L}{\partial R_{jt}} = -\lambda_{jt} + \lambda_{j,t+1} - \gamma_{jt} \le 0 \; (= 0 \text{ for } R_{jt} > 0) \\ &\lambda_{jt} \ge 0 (= 0 \text{ for } R_{jt} < R_{j,t-1} + w_{jt} - e_{jt}^H) \\ &\gamma_{jt} \ge 0 (= 0 \text{ for } R_{jt} < \bar{R}_j) \\ &\rho_{jt} \ge 0 (= 0 \text{ for } e_{jt}^H < \bar{e}_j^H) \end{aligned} \tag{10.3}$$

Let us assume that there is a positive market price in every period. The producer will not supply any electricity if his water value is higher than the market price [subtracted the shadow price on the production capacity constraint, but this shadow price is zero when production is zero according to the last condition in (10.3)]. For the periods he will supply a positive amount the market price minus the shadow price on the production capacity constraint has to be equal to his water value. This means that if the production constraint is binding, then the water value is typically lower than the market price. The producer is forced to use less water than what he wants, resulting in a forced accumulation of water or a smaller draw down than wanted. The opportunity cost of water is therefore lower than the market price.

In general the producer will strive to sell all his energy in the period with the highest price, but he may be prevented from doing this by the upper constraint on his production capacity and by threat of overflow due to the reservoir constraint. When overflow threatens his water value will be adjusted downwards for that period compared with the next period, according to the second condition in (10.3). He is willing to sell at a lower price now than a higher price in a later period to prevent overflow. But to the right price he may sell in an even earlier period and prevent an overflow situation happening. Further reasoning of a hydropower producer determining when to process his water will follow the more elaborate discussion of the model (4.8) set out in Chapter 4.

Comparing the private conditions (10.3) with the social conditions (4.12) the conditions have the same form. The only formal difference is that exogenous market prices have replaced period demand functions. If the prices faced by the producers are the same as in the social solution, and provided the planning horizon is the same for all plants and equal to the social planning horizon, then a competitive market may sustain the social solution. This is in accordance with the textbook welfare theorems in economics. But notice that we have not shown how such prices may be formed in private markets. A well functioning electricity market keeping a continuous electric equilibrium does not imply automatically that the market is also optimal in a social sense.

There are at least three problems when appealing to the welfare theorems for the type of model we are analysing. One problem is external effects created by hydrological coupled producers studied in Chapter 4. A second problem is the external effects created in a meshed network concerning losses and congestion discussed in Chapter 7. A third problem is created in the case of uncertainty. Each firm has to solve a stochastic dynamic problem, adding price uncertainty to uncertainty about own inflows. We will not treat this problem formally along the lines developed in Chapter 9, but just point out the problems created if firms operate with different price expectations. The policy of each firm will be to adjust production and reservoir level in the current period as time evolves according to the relevant expected price-and water value in the next period. With different price expectations such adaptation may create greater volatility of prices than in the social case. The firms may follow different ways of forming and updating price expectations. It is not obvious that there is a learning process leading to rational expectations since the market price is influenced by the total inflow that may be realised by many different local distributions of inflows. It is a question of what kind of information each firm has about other firms' inflow and forecasting skills.

Another problem is a possible difference in time horizon of firms. Firms with small reservoir capacity will have a shorter time horizon than firms with huge reservoir capacity. This does not necessarily lead to a deviation from the social planning solution, but may create special coordination problems that the market does not solve.

Market designs

The book has tried to establish a theoretical understanding of hydropower economics without addressing the problem of implementing a specific market structure. Starting with the deregulation in England in 1990 many different market designs have emerged (see Jamasb and Pollitt (2005) for an overview of changes within the European Union and Stoft (2002) for general considerations of market design and information about the United States). A typical feature of a market is that the wholesale market is of a day-ahead type and based on clearing hour by hour between supply and demand. In order to balance supply and demand in real time there may be a real-time market organised in advance, but the system operator may also regulate supply according to other preset arrangements with generators. Wholesale markets have by and large functioned smoothly. However, studies into the optimality of such markets in view of the social planning solution, as developed in this book, are hard to come by. National competition authorities have been focussing on use of market power (see, e.g., Report from the Nordic competition authorities, 2003), but mostly based on the distribution of market shares, which may not be the most relevant in the case of hydropower.

Problems with wholesale markets are that a greater share of trades usually occurs on a bilateral basis outside the market, and that the final consumers like households and general business are not in the market in real time. Concerning the former bilateral contracts actually may reduce problems of market power, but the question is how relevant the equilibrium price for a limited part of the market is. However, generators with bilateral contracts may profitably buy from the wholesale market if the market price is lower than the contract price, and save its own resources for periods with the opposite price relation, and big enough consumers may also buy from the wholesale market (using traders) if the price is lower on the wholesale market (provided there is no clause forcing the amount to be taken from the generator). Thus the end effect may be that wholesale market price is representative for the equilibrium price of the total volume. Consumers are represented by utilities or traders, and do not, as a rule, have real-time contracts, but price contracts of different types based on some form of *ex post* adjustment of prices. The models developed in the book are all based on demand function in real time, so there is a problem matching theoretical insights with actual market forms. There have been very limited experiments with real-time pricing, partly because measuring electricity consumption in real time is costly. The events in California 2000-2001 underlined the big problems that can arise if consumers' price is completely decoupled from the current wholesale price (Joskow and Kahn, 2002).

There is no special provision in deregulation designs for the case of hydraulically coupled hydropower stations. From Chapter 4 we saw that the most pressing coordination problem occurs when the release from an upstream plant exceeds the production capacity of the first downstream plant, and this plant is balancing a full reservoir. In a deregulated market one would expect cooperation to develop, and may be mergers of coupled plants.

The types of externalities receiving the greatest attention are the generation of loss and congestion in a network. In the economics literature emphasis has been put on potentials for use of market power playing on transmission constraints and price mark-ups on the import side of a binding constraint and price mark-downs on the export side (Hogan, 1992, 1997; Cardell et al., 1997; Bushnell, 1999). The latter price implications were also demonstrated in Chapter 6 on trade with electricity. Potential magnitudes of loss in different market systems as to incentives to deal with the loss externality in a network with loop-flows have been calculated in Green (2007) for England and Wales for 1996. The benchmark is a system with complete nodal pricing, as treated in Chapter 7, but there the physical network was not shown, thus treating transmission constraints without modelling loop-flow effects properly. As pointed out in Chapter 7 a lot of information is necessary in order for a central planner to manage a spatial pricing system in real time, and transaction costs have to be considered. However, Green (2007) comes up with impressive welfare gains if a nodal price system can be implemented. As he points out Chile, New Zealand and some regions in the United States, where PJM (Pennsylvania, New Jersey, Maryland) is the most well known, have such pricing schemes in place. When designing a nodal price system crucial decision involve the time unit and the role of prices as *ex post* device to settle account, and as *ex ante* information to generators and consumers. The PJM exchange calculates prices every five minutes for several hundred nodes. However, when it takes a conventional coal-fired thermal generator several hours both to begin producing from a cold start, at a considerable cost, and to reduce output, one may wonder about the feasibility of reacting to the price information. New Zealand has considerable hydropower, and a rather linear structure of the network not so dominated by loop-flows due to topology and location of main generators and consumer nodes. This may make it easier to implement nodal pricing as *ex ante* incentives.

Stochastic inflows were treated only within an aggregated model in Chapter 9. In the competitive wholesale market of Nord Pool several hundred independent hydro plants within the Norwegian part each has to form expectation not only about own inflows, but also about future market prices. A potential source of mismatch between the social solution and a market system is the ability to form best possible expectations. A central system operator would be most favourably placed to form expectations. The system model developed for the period of centralised coordination of the Norwegian system (Hveding, 1967-1968) has been further developed and extended to cover the Nord Pool area (Wangensteen, 2007); more or less containing the key features analysed in the book, and is used by both the regulator and by large generating companies to predict future prices. In addition, the need to hedge against uncertainty has lead to the development of futures markets at Nord Pool. The prices paid now for power deliveries weeks, months, and years ahead tell the market participants about price expectations held by market participants. These prices are, of course, public information.

Investments

The production capacities of generators, capacities of reservoirs, and capacity of the transmission network have all been assumed constant for the dynamic management problems we have addressed. Carrying out analysing optimal social investment in capacities of various types is a huge task outside the scope of this book. But calculation of the shadow prices corresponding to the given capacities will give an indication of at least the direction of desirable investment.

The shadow price on a reservoir constraint tells us the increase in the objective function of marginally increasing the reservoir capacity. This may be possible by either better utilisation of the present amount of water by reducing friction inside tunnels, increasing the size of the reservoir, or by increasing the catchments of water into previously untouched sources.

The costs of such investments can be calculated. The point is now that the benefit side of a marginal investment is the sum of the positive shadow prices within the horizon. It may not be feasible to carry out a marginal investment, but this simple cost-benefit calculation gives an indication of whether it is interesting to carry out investment analyses. In a system characterised by optimal amount of capacity there should be equality between benefit and costs at the margin, provided sufficient flexibility of dimensioning the investment project.

Whether production capacity should be increased can be investigated by a similar comparison of the sum of positive shadow prices and the cost of investment. If the turbine capacity is the limiting factor the investment project is not so large, but if the water-feeding capacity through tunnels from reservoirs has to be increased, this is a more major undertaking.

Shadow prices on the environmental constraints introduced in Chapter 4 can serve as a basis for discussing the rationale of the constraints. Environmental costs (or benefit of current regulation) should be quantified and compared with the shadow value of marginally relaxing the constraints. The result of such calculations may work both ways as to which way to change. If water-flows downstream of power plants are based on the need to transport timber in the timber-floating season, this does not make much sense years after lorries have taken over such transports. On the other hand, demand for river-based recreation of various types, or willingness to pay for unspoiled ecosystems of rivers, may have increased considerably.

Investments in networks are of special theoretical interest because of the loss and congestion externalities present, as expanded upon in Chapter 7. It is rather obvious that investment in lines for given production will not only reduce loss, but will then necessarily contribute to increased consumption. The analogy is with the best investment of a waterworks may be to reduce leakages instead of expanding to new water sources. Within the framework in Chapter 7, using individual thermal constraints for lines and not modelling loop-flows, the numerical values of the shadow prices of binding line constraints may give some useful information for investment decisions.

An additional benefit of "over-investing" in transmission capacity is the effect on reducing the possibility of using market power by creating isolated electricity areas manipulating congestion of lines.

Shift in demand over time for electricity necessitates investments both in generating capacity and in transmission capacity. These investments cannot be carried out in isolation, but owing to loss and congestion externalities have to be considered simultaneously in order to achieve optimal social return on the investments.

Returning to deregulation of electricity markets the Nord Pool area has experienced a markedly lack of investments of both types the last decennium. This may be due to earlier over-investments, but may also reflect uncertainties involved, and private investors still waiting for a high enough trigger price of electricity so the option value of investments also gets covered. The role of the incentive effects of the market design seems to be an interesting topic for future research.

References

Ambec, S. and Doucet, J. A. (2003): "Decentralizing hydro power production," *Canadian Journal of Economics* 36, 587-607.

Baumol, W. J. (1972): *Economic theory and operations analysis*, London: Prentice/Hall International, Inc. Third edition.

Bellman, R. (1957): *Dynamic programming*, Princeton: Princeton University Press.

Bohn, R. E., Caramanis, M. C. and Schweppe, F. C. (1984): "Optimal pricing in electrical networks over space and time," *Rand Journal of Economics* 15(3), 360-376.

Borenstein, S., Bushnell, J. B. and Wolak, F. A. (2002): "Measuring market inefficiencies in California's restructured wholesale electricity market," *American Economic Review* 92(5), 1376-1405.

Borenstein, S., Bushnell, J. B. and Knittel, C. R. (1999): "Market power in electricity markets: beyond concentration measures," *Energy Journal* 20(4), 65-88.

Borenstein, S., Bushnell, J. B., and Stoft, S. (2000): "The competitive effects of transmission capacity in a deregulated electricity market," *Rand Journal of Economics* 31(2), 294-325.

Bushnell, J. B. (1999): "Transmission rights and market power," *Electricity Journal* 12(8), 77-85.

Bushnell, J. (2003): "A mixed complementarity model of hydrothermal electricity competition in the Western United States," *Operations Research* 2003, 51(1) January-February.

Cardell, J. B., Hitt, C. C. and Hogan, W. W. (1997): "Market power model and strategic interaction in electricity networks," *Resource and Energy Economics* 19, 109-137.

Crampes, C. and Moreaux, M. (2001): "Water resource and power generation," *International Journal of Industrial Organization* 19, 975-997.

Edwards, B. K. (2003): *The economics of hydroelectric power*, Cheltenham / Northampton: Edward Elgar.

von der Fehr, N.-H. M. and Sandsbråten, L. (1997): "Water on fire: gains from electricity trade," *Scandinavian Journal of Economics* 99, 281-297.

Frisch, R. (1965): *Theory of production,* Dordrecht: D. Reidel.

Førsund, F. R. (1994): "Driftsoptimalisering i vannkraftsystemet" [Operation optimisation in the hydropower system], SNF report 29/94, SNF, Oslo.

Førsund, F. R. (1999): "On the contribution of Ragnar Frisch to production theory," *Rivista Internazionale di Scienze Economiche e Commerciali* (*International Review of Economics and Business*) XLVI, 1-34.

Førsund, F. R. and Hjalmarsson, L. (1987): *Analyses of Industrial Structure: A Putty-Clay Approach,* The Industrial Institute for Economic and Social Research, Stockholm: Almqvist and Wiksell International.

Garcia, A., Reitzes, J.D. and Stacchetti, E. (2001): "Strategic pricing when electricity is storable," *Journal of Regulatory Economics* 20(3), 223-247.

Gessford, J. and Karlin, S. (1958): "Optimal policy for hydroelectric operation," in Arrow, K.J., Karlin, S., Scarf, H. (eds.) *Studies in the Mathematical Theory of Inventory and Production*, Stanford: Stanford University Press, 179-200.

Gjelsvik, A., Røtting, T. A. and Røynstrand, J. (1992): "Long-term scheduling of hydro-thermal power systems," in Broch and Lysne (eds): *Hydropower 92*, Rotterdam: Balkema, 539-546.

Green, R. J. (2007): "Nodal pricing of electricity: how much does it cost to get it wrong?," *Journal of Regulatory Economics* 31, 125-149.

Green, R. J. and Newbery, D. M. (1992): "Competition in the British electricity spot market," *Journal of Political Economy* 100(5), 929-953.

Hansen, P. V. (2007): "Inflow uncertainty in hydropower markets," forthcoming in *Discussion Papers*, Statistics Norway.

Haugstad, A., Reinertsen, T. Green, T. Johannesen, A. and Botnen, O. J. (1990): "Samkjøringsmodellen. Et verktøy for drifts- og utbyggingsplanlegging" [The coordination model. A tool for operation- and construction planning], EFI TR A3678.

Hogan, W. W. (1992): "Contract networks for electric power transmission," *Journal of Regulatory Economics* 4(3), 211-242.

Hogan, W. W. (1997): "A market power model with strategic interaction in electricity networks," *Energy Journal* 18(4), 107-141.

Hsu, M. (1997): "An introduction to the pricing of electric power transmission," *Utilities Policy* 6(3), 257-270.

Hveding, V. (1967): "Driftssimulering av et kraftproduksjonssystem" [Operations simulation of a power-production system], *Elektroteknisk Tidsskrift* 80 (25), 517-525.

Hveding, V. (1968): "Digital simulation techniques in power system planning," *Economics of Planning* 8(2), 118-139.

Jamasb, T. and Pollitt, M. (2005): "Electricity reforms in the European Union: review of progress toward liberalisation and integration," *Energy Journal* (*Special issue, European Electricity Liberalisation*) 26, 11-42.

Johansen, L. (1972): *Production Functions. An integration of micro and macro, short run and long run aspects*, Amsterdam: North-Holland.

Johannesen, A. and Flatabø, N. (1989): "Scheduling methods in operation planning of a hydro-dominated power production system," *Electrical Power & Energy Systems* 11(3), 189-199.

Johnsen, T. A. (2001): "Hydropower generation and storage, transmission constraints and market power," *Utilities Policy* 10, 63-73.

Joskow, P. L. and Kahn, E. (2002): "A quantitative analysis of pricing behavior in California's wholesale electricity market during summer 2000," *Energy Journal* 23(4), 1-35.

Kelman, R., Barroso, L. A. N. and Pereira, M. V. F. (2001): "Market power assessment and mitigation in hydrothermal systems," *IEEE Transactions on Power Systems* 16(3), 354-359.

Koopmans, T. C. (1957): "Water storage policy in a simplified hydroelectric system," *Proceedings of the First International Conference on Operational Research*, Oxford, September, 1-35.

Larsen, B. M. and Nesbakken, R. (2005): "Formålsfordeling av husholdningenes elektrisitetsforbruk i 2001" [Purpose distribution of households' electricity consumption in 2001], Reports 2005/18, Statistics Norway, Oslo – Kongsvinger.

Lien, H. G. (2006): "Ressursbruk i vannkraftproduksjon" [Resource use in hydropower production], Master Thesis, Department of Economics, University of Oslo.

Little, J. D. C. (1955): "The use of storage water in a hydroelectric system," *Journal of the Operations Research Society of America* 3, 187-197.

Massé, P. B. D. (1946): *Les reserves et la regulation de l'avenir dans la vie économique* [*Reservoirs and regulation for future economic activity*], Paris: Herman.

Morlat, G. (1964): "On instructions for the optimum management of seasonal reservoirs," in Nelson, J. R. (ed.): *Marginal cost pricing in practice*, Englewood Cliffs: Prentice-Hall Inc., 159-179.

Newbery, D. M. (2005): "Electricity liberalisation in Britain: the quest for a satisfactory wholesale market design," *Energy Journal* (*Special issue, European Electricity Liberalisation*) 26, 43-70.

NOU (1979): *Kraftintensiv industri* [*Power-intensive industry*], Norwegian Official Reports (NOU), 1979:49, Oslo, Bergen, Tromsø: Universitetsforlaget.

OED (2005): *Fakta 2005 om energi- og vassdragsvirksomhet* [*Facts 2005 on energy- and waterway activities*], Oslo: Norwegian Ministry of Petroleum and Energy (OED).

Pereira, M. V. F. (1989): "Optimal stochastic operations scheduling of large hydroelectric systems," *Electrical Power & Energy Systems* 11(3), 161-169.

Pereira, M. V. F. and Pinto, L. M. V. G. (1985): "Stochastic optimization of a multireservoir hydroelectric system: a decomposition approach," *Water Resources Research* 21(6), 779-792.

Pereira, M. V. F. and Pinto, L. M. V. G. (1991): "Multi-stage stochastic optimization applied to energy planning," *Mathematical Programming* 52, 359-375.

Report from the Nordic competition authorities (2003): "A powerful competition policy," Report No. 1/2003.

Rothchild, M. and Stiglitz, J. E. (1970): "Increasing risk: 1. A definition," *Journal of Economic Theory* 2, 225-243.

Scott, T. J. and Read, E. G. (1996): "Modelling hydro reservoir operation in a deregulated electricity market," *International Transactions in Operational Research* 3(3/4), 243-253.

Schweppe, F. C., Caramanis, M. C., Tabors, R. D. and Bohn, R. E. (1988): *Spot pricing of electricity*, Boston/Dordrecht/London: Kluwer Academic Publishers.

Smith, V. L. (1961): *Investment and production*, Cambridge: Harvard University Press.

Stage, S. and Larsson, Y. (1961): "Incremental cost of water power," *AIEE Transactions, Part III (Power Apparatus and Systems)* 80, 361-365.

Stoft, S. (2002): *Power system economics*, Piscataway: IEEE Press, Wiley-Interscience, John Wiley & Sons, Inc.

Sydsæter, K., Strøm, A. and Berck, P. (1999): *Economists' mathematical manual*, Berlin, Heidelberg, New York: Springer Verlag.

Sydsæter, K., Hammond, P., Seierstad, A. and Strøm, A. (2005): *Further mathematics for economic analysis*, Harlow: Pearson Education Limited.

Thomson, W. (1881): "On the economy of metal in conductors of electricity," in *Report of the British Association for the Advancement of Science, Sec. A*, September 2, 526-528, London: John Murray.

Vickrey, W. (1971): "Responsive pricing of public utility services," *Bell Journal of Economics and Management Science* 2, 337-346.

Wangensteen, I. (2007): *Power system economics – the Nordic electricity market*, Trondheim: Tapir Academic Press.

Wallace, S. W. and Fleten, S.-E. (2002): "Stochastic programming models in energy," in: A. Ruszczynski and A. Shapiro, eds., *Handbooks in Operations Research and Management Science* 10, North – Holland.

Willoch, K. (1985): "El-forsyningen foran store oppgaver" [Electricity supply faces great challenges], *Fossekallen* 31(7), 4-5.

Wood, A. J. and Wollenberg, B. F. (1984): *Power generation, operation, and control*, New York: John Wiley & Sons, Inc., Second edition, 1996.

Author Index

Subject Index

Early Titles in the
INTERNATIONAL SERIES IN
OPERATIONS RESEARCH & MANAGEMENT SCIENCE
Frederick S. Hillier, Series Editor, *Stanford University*

Saigal/ *A MODERN APPROACH TO LINEAR PROGRAMMING*
Nagurney/ *PROJECTED DYNAMICAL SYSTEMS & VARIATIONAL INEQUALITIES WITH APPLICATIONS*
Padberg & Rijal/ *LOCATION, SCHEDULING, DESIGN AND INTEGER PROGRAMMING*
Vanderbei/ *LINEAR PROGRAMMING*
Jaiswal/ *MILITARY OPERATIONS RESEARCH*
Gal & Greenberg/ *ADVANCES IN SENSITIVITY ANALYSIS & PARAMETRIC PROGRAMMING*
Prabhu/ *FOUNDATIONS OF QUEUEING THEORY*
Fang, Rajasekera & Tsao/ *ENTROPY OPTIMIZATION & MATHEMATICAL PROGRAMMING*
Yu/ *OR IN THE AIRLINE INDUSTRY*
Ho & Tang/ *PRODUCT VARIETY MANAGEMENT*
El-Taha & Stidham/ *SAMPLE-PATH ANALYSIS OF QUEUEING SYSTEMS*
Miettinen/ *NONLINEAR MULTIOBJECTIVE OPTIMIZATION*
Chao & Huntington/ *DESIGNING COMPETITIVE ELECTRICITY MARKETS*
Weglarz/ *PROJECT SCHEDULING: RECENT TRENDS & RESULTS*
Sahin & Polatoglu/ *QUALITY, WARRANTY AND PREVENTIVE MAINTENANCE*
Tavares/ *ADVANCES MODELS FOR PROJECT MANAGEMENT*
Tayur, Ganeshan & Magazine/ *QUANTITATIVE MODELS FOR SUPPLY CHAIN MANAGEMENT*
Weyant, J. / *ENERGY AND ENVIRONMENTAL POLICY MODELING*
Shanthikumar, J.G. & Sumita, U./ *APPLIED PROBABILITY AND STOCHASTIC PROCESSES*
Liu, B. & Esogbue, A.O./ *DECISION CRITERIA AND OPTIMAL INVENTORY PROCESSES*
Gal, T., Stewart, T.J., Hanne, T. / *MULTICRITERIA DECISION MAKING: Advances in MCDM Models, Algorithms, Theory, and Applications*
Fox, B.L. / *STRATEGIES FOR QUASI-MONTE CARLO*
Hall, R.W. / *HANDBOOK OF TRANSPORTATION SCIENCE*
Grassman, W.K. / *COMPUTATIONAL PROBABILITY*
Pomerol, J-C. & Barba-Romero, S. / *MULTICRITERION DECISION IN MANAGEMENT*
Axsäter, S. / *INVENTORY CONTROL*
Wolkowicz, H., Saigal, R., & Vandenberghe, L. / *HANDBOOK OF SEMI-DEFINITE PROGRAMMING: Theory, Algorithms, and Applications*
Hobbs, B.F. & Meier, P. / *ENERGY DECISIONS AND THE ENVIRONMENT: A Guide to the Use of Multicriteria Methods*
Dar-El, E. / *HUMAN LEARNING: From Learning Curves to Learning Organizations*
Armstrong, J.S. / *PRINCIPLES OF FORECASTING: A Handbook for Researchers and Practitioners*
Balsamo, S., Personé, V., & Onvural, R./ *ANALYSIS OF QUEUEING NETWORKS WITH BLOCKING*
Bouyssou, D. et al. / *EVALUATION AND DECISION MODELS: A Critical Perspective*
Hanne, T. / *INTELLIGENT STRATEGIES FOR META MULTIPLE CRITERIA DECISION MAKING*
Saaty, T. & Vargas, L. / *MODELS, METHODS, CONCEPTS and APPLICATIONS OF THE ANALYTIC HIERARCHY PROCESS*
Chatterjee, K. & Samuelson, W. / *GAME THEORY AND BUSINESS APPLICATIONS*
Hobbs, B. et al. / *THE NEXT GENERATION OF ELECTRIC POWER UNIT COMMITMENT MODELS*
Vanderbei, R.J. / *LINEAR PROGRAMMING: Foundations and Extensions, 2nd Ed.*
Kimms, A. / *MATHEMATICAL PROGRAMMING AND FINANCIAL OBJECTIVES FOR SCHEDULING PROJECTS*
Baptiste, P., Le Pape, C. & Nuijten, W. / *CONSTRAINT-BASED SCHEDULING*
Feinberg, E. & Shwartz, A. / *HANDBOOK OF MARKOV DECISION PROCESSES: Methods and Applications*
Ramík, J. & Vlach, M. / *GENERALIZED CONCAVITY IN FUZZY OPTIMIZATION AND DECISION ANALYSIS*
Song, J. & Yao, D. / *SUPPLY CHAIN STRUCTURES: Coordination, Information and Optimization*
Kozan, E. & Ohuchi, A. / *OPERATIONS RESEARCH/ MANAGEMENT SCIENCE AT WORK*

Early Titles in the
INTERNATIONAL SERIES IN
OPERATIONS RESEARCH & MANAGEMENT SCIENCE
(Continued)

Bouyssou et al. / *AIDING DECISIONS WITH MULTIPLE CRITERIA: Essays in Honor of Bernard Roy*
Cox, Louis Anthony, Jr. / *RISK ANALYSIS: Foundations, Models and Methods*
Dror, M., L'Ecuyer, P. & Szidarovszky, F. / *MODELING UNCERTAINTY: An Examination of Stochastic Theory, Methods, and Applications*
Dokuchaev, N. / *DYNAMIC PORTFOLIO STRATEGIES: Quantitative Methods and Empirical Rules for Incomplete Information*
Sarker, R., Mohammadian, M. & Yao, X. / *EVOLUTIONARY OPTIMIZATION*
Demeulemeester, R. & Herroelen, W. / *PROJECT SCHEDULING: A Research Handbook*
Gazis, D.C. / *TRAFFIC THEORY*
Zhu/ *QUANTITATIVE MODELS FOR PERFORMANCE EVALUATION AND BENCHMARKING*
Ehrgott & Gandibleux/*MULTIPLE CRITERIA OPTIMIZATION: State of the Art Annotated Bibliographical Surveys*
Bienstock/ *Potential Function Methods for Approx. Solving Linear Programming Problems*
Matsatsinis & Siskos/ *INTELLIGENT SUPPORT SYSTEMS FOR MARKETING DECISIONS*
Alpern & Gal/ *THE THEORY OF SEARCH GAMES AND RENDEZVOUS*
Hall/*HANDBOOK OF TRANSPORTATION SCIENCE - 2nd Ed.*
Glover & Kochenberger/ *HANDBOOK OF METAHEURISTICS*
Graves & Ringuest/ *MODELS AND METHODS FOR PROJECT SELECTION: Concepts from Management Science, Finance and Information Technology*
Hassin & Haviv/ *TO QUEUE OR NOT TO QUEUE: Equilibrium Behavior in Queueing Systems*
Gershwin et al. /*ANALYSIS & MODELING OF MANUFACTURING SYSTEMS*
Maros/ *COMPUTATIONAL TECHNIQUES OF THE SIMPLEX METHOD*
Harrison, Lee & Neale/ *THE PRACTICE OF SUPPLY CHAIN MANAGEMENT: Where Theory and Application Converge*
Shanthikumar, Yao & Zijm/ *STOCHASTIC MODELING AND OPTIMIZATION OF MANUFACTURING SYSTEMS AND SUPPLY CHAINS*
Nabrzyski, Schopf & Węglarz/ *GRID RESOURCE MANAGEMENT: State of the Art and Future Trends*
Thissen & Herder/ *CRITICAL INFRASTRUCTURES: State of the Art in Research and Application*
Carlsson, Fedrizzi, & Fullér/ *FUZZY LOGIC IN MANAGEMENT*
Soyer, Mazzuchi & Singpurwalla/ *MATHEMATICAL RELIABILITY: An Expository Perspective*
Chakravarty & Eliashberg/ *MANAGING BUSINESS INTERFACES: Marketing, Engineering, and Manufacturing Perspectives*
Talluri & van Ryzin/ *THE THEORY AND PRACTICE OF REVENUE MANAGEMENT*
Kavadias & Loch/*PROJECT SELECTION UNDER UNCERTAINTY: Dynamically Allocating Resources to Maximize Value*
Brandeau, Sainfort & Pierskalla/ *OPERATIONS RESEARCH AND HEALTH CARE: A Handbook of Methods and Applications*
Cooper, Seiford & Zhu/ *HANDBOOK OF DATA ENVELOPMENT ANALYSIS: Models and Methods*
Luenberger/ *LINEAR AND NONLINEAR PROGRAMMING, 2nd Ed.*
Sherbrooke/ *OPTIMAL INVENTORY MODELING OF SYSTEMS: Multi-Echelon Techniques, Second Edition*
Chu, Leung, Hui & Cheung/ *4th PARTY CYBER LOGISTICS FOR AIR CARGO*
Simchi-Levi, Wu & Shen/ *HANDBOOK OF QUANTITATIVE SUPPLY CHAIN ANALYSIS: Modeling in the E-Business Era*
Gass & Assad/ *AN ANNOTATED TIMELINE OF OPERATIONS RESEARCH: An Informal History*
Greenberg/ *TUTORIALS ON EMERGING METHODOLOGIES AND APPLICATIONS IN OPERATIONS RESEARCH*

Early Titles in the
INTERNATIONAL SERIES IN
OPERATIONS RESEARCH & MANAGEMENT SCIENCE
(Continued)

Weber/ *UNCERTAINTY IN THE ELECTRIC POWER INDUSTRY: Methods and Models for Decision Support*
Figueira, Greco & Ehrgott/ *MULTIPLE CRITERIA DECISION ANALYSIS: State of the Art Surveys*
Reveliotis/ *REAL-TIME MANAGEMENT OF RESOURCE ALLOCATIONS SYSTEMS: A Discrete Event Systems Approach*
Kall & Mayer/ *STOCHASTIC LINEAR PROGRAMMING: Models, Theory, and Computation*

***** *A list of the more recent publications in the series is at the front of the book* *****

FSC
www.fsc.org
MIX
Papier | Fördert
gute Waldnutzung
FSC® C083411

Kolibri 360 GmbH
Ferdinand-Jühlke-Straße 7
99095 Erfurt, Deutschland
produktsicherheit@kolibri360.de